OECD/G20 Base Erosion and Profit Shifting Project

Neutralising the Effects of Branch Mismatch Arrangements, Action 2

INCLUSIVE FRAMEWORK ON BEPS

This document and any map included herein are without prejudice to the status of or sovereignty over any territory, to the delimitation of international frontiers and boundaries and to the name of any territory, city or area.

Please cite this publication as:
OECD (2017), *Neutralising the Effects of Branch Mismatch Arrangements, Action 2: Inclusive Framework on BEPS*, OECD/G20 Base Erosion and Profit Shifting Project, OECD Publishing, Paris.
http://dx.doi.org/10.1787/9789264278790-en

ISBN 978-92-64-27795-3 (print)
ISBN 978-92-64-27879-0 (PDF)

Series: OECD/G20 Base Erosion and Profit Shifting Project
ISSN 2313-2604 (print)
ISSN 2313-2612 (online)

The statistical data for Israel are supplied by and under the responsibility of the relevant Israeli authorities. The use of such data by the OECD is without prejudice to the status of the Golan Heights, East Jerusalem and Israeli settlements in the West Bank under the terms of international law.

Photo credits: Cover © ninog – Fotolia.com

Corrigenda to OECD publications may be found on line at: *www.oecd.org/about/publishing/corrigenda.htm*.
© OECD 2017

You can copy, download or print OECD content for your own use, and you can include excerpts from OECD publications, databases and multimedia products in your own documents, presentations, blogs, websites and teaching materials, provided that suitable acknowledgement of OECD as source and copyright owner is given. All requests for public or commercial use and translation rights should be submitted to *rights@oecd.org*. Requests for permission to photocopy portions of this material for public or commercial use shall be addressed directly to the Copyright Clearance Center (CCC) at *info@copyright.com* or the Centre français d'exploitation du droit de copie (CFC) at *contact@cfcopies.com*.

Foreword

The integration of national economies and markets has increased substantially in recent years, putting a strain on the international tax rules, which were designed more than a century ago. Weaknesses in the current rules create opportunities for base erosion and profit shifting (BEPS), requiring bold moves by policy makers to restore confidence in the system and ensure that profits are taxed where economic activities take place and value is created.

Following the release of the report *Addressing Base Erosion and Profit Shifting* in February 2013, OECD and G20 countries adopted a 15-point Action Plan to address BEPS in September 2013. The Action Plan identified 15 actions along three key pillars: introducing coherence in the domestic rules that affect cross-border activities, reinforcing substance requirements in the existing international standards, and improving transparency as well as certainty.

After two years of work, measures in response to the 15 actions were delivered to G20 Leaders in Antalya in November 2015. All the different outputs, including those delivered in an interim form in 2014, were consolidated into a comprehensive package. The BEPS package of measures represents the first substantial renovation of the international tax rules in almost a century. Once the new measures become applicable, it is expected that profits will be reported where the economic activities that generate them are carried out and where value is created. BEPS planning strategies that rely on outdated rules or on poorly co-ordinated domestic measures will be rendered ineffective.

Implementation is now the focus of this work. The BEPS package is designed to be implemented via changes in domestic law and practices, and via treaty provisions. With the negotiation for a multilateral instrument (MLI) having been finalised in 2016 to facilitate the implementation of the treaty related measures, 67 countries signed the MLI on 7 June 2017, paving the way for swift implementation of the treaty related measures. OECD and G20 countries also agreed to continue to work together to ensure a consistent and co-ordinated implementation of the BEPS recommendations and to make the project more inclusive. Globalisation requires that global solutions and a global dialogue be established which go beyond OECD and G20 countries.

As a result, the OECD established an Inclusive Framework on BEPS, bringing all interested and committed countries and jurisdictions on an equal footing in the Committee on Fiscal Affairs and all its subsidiary bodies. The Inclusive Framework, which already has 100 members, will monitor and peer review the implementation of the minimum standards as well as complete the work on standard setting to address BEPS issues. In addition to BEPS Members, other international organisations and regional tax bodies are involved in the work of the Inclusive Framework, which also consults business and the civil society on its different work streams.

A better understanding of how the BEPS recommendations are implemented in practice could reduce misunderstandings and disputes between governments. Greater focus on implementation and tax administration should therefore be mutually beneficial to governments and business. Proposed improvements to data and analysis will help support ongoing evaluation of the quantitative impact of BEPS, as well as evaluating the impact of the countermeasures developed under the BEPS Project.

Table of contents

Abbreviations and acronyms 7

Executive summary 9

Introduction 13
    Branch payee structures that give rise to D/NI outcomes 13
    Deemed branch payments 16
    DD branch payments 17
    Imported branch mismatches 18
    Summary of Recommendations 19

Chapter 1. **Limitation to the scope of the branch exemption** 23
    Overview 24
    Recommendation 1.1 – Limitation to the scope of the branch exemption .. 25

Chapter 2. **Branch payee mismatch rule** 27
    Overview 28
    Recommendation 2.1 – Denial of deduction for branch payee mismatches .. 28
    Recommendation 2.2 – Disregarded branch 31
    Recommendation 2.3 – Scope of the rule 32

Chapter 3. **Deemed branch payment rule** 35
    Overview 36
    Recommendation 3.1 – Denial of deduction for deemed branch payments .. 36
    Recommendation 3.2 – Deemed branch payments 37
    Recommendation 3.3 – Rule only applies to payments that result in a branch mismatch .. 40

Chapter 4. **Double Deduction Rule** 43
    Overview 44
    Recommendation 4.1 – Treatment of DD outcomes 44
    Recommendation 4.2 – DD outcome 44
    Recommendation 4.3 – No branch mismatch to the extent set off against dual inclusion income 46

Chapter 5. **Imported branch mismatch rule** 49
    Overview 50
    Recommendation 5.1 – Treatment of imported branch mismatches 50
    Recommendation 5.2 – Imported branch mismatch definition 51
    Recommendation 5.3 – Limitations on scope 52

Annex A. Summary of recommendations 53

Annex B. List of examples 57

Figures

Figure 1	Disregarded branch structure	14
Figure 2	Diverted branch payment	15
Figure 3	Deemed branch payment	16
Figure 4	DD branch payment	18
Figure 5	Imported branch mismatches	19

Tables

Table B.2.1	Taxable Branch	63
Table B.3.1	Taxable Branch with non-dual inclusion income	65
Table B.4.1	Exempt branch	68
Table B.4.2	Exempt branch recognising deemed payment in payee jurisdiction	69
Table B.4.3	Exempt branch with deduction for branch income	70
Table B.4.4	Adjustment under Recommendation 3 for Exempt Branch	72
Table B.5.1	Mismatch arising in respect of deemed and actual payments	74
Table B.5.2	Adjustment under Recommendation 3	76
Table B.5.3	Calculation of total deductions claimed in Branch and Head Office	77
Table B.5.4	Adjustments under Recommendations 3 and 4	78
Table B.6.1	Mismatch arising in respect of deemed and actual payments	80
Table B.6.2	Calculation of total deductions claimed in each jurisdiction	81
Table B.6.3	Adjustments under Recommendations 3 and 4	82
Table B.7.1	Calculation of expenditure in each jurisdiction	84
Table B.7.2	Adjustments under Country B and C law	85
Table B.8.1	Mismatch arising from notional payment	87
Table B.8.2	Adjustment under Recommendation 3	89
Table B.9.1	Mismatch arising from double deduction	91
Table B.9.2	Adjustment under Recommendation 4.1 (a)	92
Table B.9.3	Adjustment under Recommendation 4.1 (b)	92
Table B.10.1	Net income (and loss) positions	94
Table B.10.2	Expected tax outcomes in Country A	95
Table B.10.3	Expected tax outcomes in Country B	95
Table B.10.4	Adjustment under Recommendation 4	96
Table B.10.5	Adjustment under Recommendation 4	97

Abbreviations and acronyms

ATAD	Anti-Tax Avoidance Directive
BEPS	Base Erosion and Profit Shifting
CFA	Committee on Fiscal Affairs
CFC	Controlled Foreign Company
DD	Double deduction
D/NI	Deduction/no inclusion
IP	Intellectual Property
OECD	Organisation for Economic Co-operation and Development
PE	Permanent Establishment
R&D	Research and Development
SPV	Special Purpose lending Vehicle
WP11	Working Party No.11 on Aggressive Tax Planning

Executive summary

The Report on *Neutralising the Effects of Hybrid Mismatch Arrangements* (Action 2 Report, OECD 2015) sets out recommendations for domestic rules designed to neutralise mismatches in tax outcomes that arise in respect of payments under a hybrid mismatch arrangement. The recommendations in Chapters 3 to 8 of that report set out rules targeting payments made by or to a hybrid entity that give rise to one of three types of mismatches:

a. deduction/no inclusion (D/NI) outcomes, where the payment is deductible under the rules of the payer jurisdiction but not included in the ordinary income of the payee

b. double deduction (DD) outcomes, where the payment triggers two deductions in respect of the same payment

c. indirect deduction/no inclusion (indirect D/NI) outcomes, where the income from a deductible payment is set off by the payee against a deduction under a hybrid mismatch arrangement.

The Action 2 Report (OECD 2015) includes specific recommendations for improvements to domestic law intended to reduce the frequency of such mismatches as well as targeted hybrid mismatch rules which adjust the tax consequences in either the payer or payee jurisdiction in order to neutralise the hybrid mismatch without disturbing any of the other tax, commercial or regulatory outcomes.

The Action 2 Report considers mismatches that are the result of differences in the tax treatment or characterisation of an instrument or entity. The report does not directly consider similar issues that can arise through the use of branch structures. These branch mismatches occur where the residence jurisdiction (i.e. the jurisdiction in which the head office is established) and a branch jurisdiction (i.e. the jurisdiction in which the branch is located) take a different view as to the allocation of income and expenditure between the branch and head office and include situations where the branch jurisdiction does not treat the taxpayer as having a taxable presence in that jurisdiction.

Branch mismatches are a product of inconsistencies in the domestic rules for determining the amount of income and expenditure subject to tax in each jurisdiction where the taxpayer operates. Branch mismatches exploit both differences in the domestic rules for determining whether an enterprise is subject to tax in a particular jurisdiction and the amount of income and expenditure to be taken into account in calculating that tax liability. For example, the residence jurisdiction may include all of the taxpayer's income on a worldwide basis (including all the income of foreign branches) while providing the taxpayer with a tax credit or exemption to eliminate double taxation on foreign income, while the branch jurisdiction treats the branch operation as a separate enterprise and taxes only the net income properly attributable to the branch. Although both these approaches to calculating the net income of the taxpayer in each jurisdiction may be intended to ensure that the taxpayer's entire net income is subject to tax in at least one jurisdiction (while avoiding economic double taxation of the same income), the different approaches to calculating that income may allow the taxpayer to leave an item of income out of the charge to taxation or allow the same item of expenditure to be deducted twice from the net income in two jurisdictions. Alternatively, the effect of an adjustment in one jurisdiction may be ignored in the other, thereby reducing the aggregate amount of income that the taxpayer is required to bring into charge to taxation.

Branch mismatch arrangements can be used to produce the same types of mismatches that are targeted by the recommendations in the Action 2 Report (OECD, 2015). For example:

a. A deductible payment made to a branch may not be brought into income in either the branch or residence jurisdiction (a D/NI outcome analogous to that described in Chapters 4 and 5 of the Action 2 Report (OECD, 2015)).

b. A branch may make (or be treated as making) a deductible payment to the head office that is not taken into account in calculating the net income of the head office under the laws of the residence jurisdiction (a D/NI outcome analogous to that described in Chapter 3 of the Action 2 Report (OECD, 2015)).

c. The same item of expenditure may be treated as deductible under the laws of both the residence and the branch jurisdictions (a DD outcome analogous to that described in Chapters 6 and 7 of the Action 2 Report (OECD, 2015)).

d. The income from a payment may be offset against a deduction under a branch mismatch arrangement (an indirect D/NI outcome analogous to that described in Chapter 8 of the Action 2 Report (OECD, 2015)).

Branch mismatch arrangements offer multinationals opportunities to reduce their overall tax burden by exploiting differences in the rules governing the allocation of payments between two jurisdictions, thereby raising the same issues as hybrid mismatches in terms of competition, transparency, efficiency and fairness. While a taxpayer's decision to operate through a branch will generally be driven by commercial or regulatory (rather than tax) factors, the mismatch that arises under the branch structure is the result of a taxpayer exploiting inconsistent positions adopted by the residence and branch jurisdiction on the allocation of income and expenditure between the branch and head office. For example, in the case of diverted branch payments, the mismatch arises due to the fact that the payee does not take a payment into account in either the residence or the branch jurisdiction. In the case of double deduction structures, the taxpayer deducts the same expense in different jurisdictions and sets that deduction off against income that is not subject to tax in the other jurisdiction while, in the case of a deemed branch payment, the payer is generally compensating the payee for an asset, function or risk that the payee does not treat itself as holding, performing or bearing for tax purposes.

Mismatches will not arise where all jurisdictions adhere to a common standard in the rules for determining a taxable presence and in the allocation of income or expenditure to different parts of the same enterprise and those standards are applied consistently by the taxpayer in both jurisdictions. Such international standards are the primary solution for addressing such mismatches. A number of the BEPS Action Items set out modifications to international tax standards that may reduce the BEPS opportunities associated with these types of mismatches. For example:

a. The Action 7 Report on *Preventing the Artificial Avoidance of Permanent Establishment Status* (OECD, 2015) includes recommendations for changes to the permanent establishment definition to address techniques used to inappropriately avoid creating a taxable presence in the branch jurisdiction.

b. The Report on Actions 8-10 (*Aligning Transfer Pricing outcomes with Value Creation* (OECD 2015)) sets out changes to the transfer pricing guidelines designed to ensure that the transfer pricing of multinational enterprises better aligns the taxation of profits with economic activity.

In practice, however, differences between the rules (or in the application of the rules) for calculating the net income of a branch or head office will continue to exist in those cases where both jurisdictions have not aligned their rules and practice in accordance with a common standard. While, the most comprehensive and effective way of addressing differences in the allocation of profit between the branch and head office would be for all jurisdictions to adhere to a single standard in attributing and calculating branch income, in the absence of this type of harmonisation, a country cannot protect its tax base from the risks posed by branch mismatches simply by adhering to such an agreed standard. The recommendations set out in this report call for one-off adjustments in order to neutralise tax planning opportunities that can arise in those cases where taxpayers exploit differences in the methodology for calculating the net income of the branch and head office.

Given the similarity between hybrid and branch mismatches, both in terms of structure and outcomes, countries that have adopted hybrid mismatch rules have, at the same time, generally chosen to adopt an equivalent and parallel set of rules targeting branch mismatches.[1] These branch mismatch rules apply the same analysis and solutions set out in the Action 2 Report (OECD, 2015) to neutralise mismatches that arise in the branch context. The adoption of branch and hybrid mismatch rules as a single package supports the integrity of the common approach set out in Action 2 by aligning the treatment of both types of mismatches and thereby preventing taxpayers shifting from hybrid mismatch to branch mismatch arrangements in order to secure the same tax advantages.

On 22 August 2016, the Committee on Fiscal Affairs (CFA) issued a discussion document on branch mismatch arrangements[2] inviting interested parties to comment on recommendations for branch mismatch rules that would bring the treatment of these structures into line with the treatment of hybrid mismatch arrangements as set out in the Action 2 Report (OECD, 2015). The recommendations in this report have been prepared in light of the comments received on that discussion document and the legal changes that countries have made since the release of the Action 2 Report (OECD, 2015).

The introduction to this report describes the various categories of branch mismatch arrangement covered by this report and the recommendations for specific changes to domestic law and branch mismatch rules that would bring the tax treatment of these arrangements into line with the common approach set out in the Action 2 Report (OECD, 2015), are set out in Chapters 1-5.

Annex A of this report summarises the recommendations and Annex B sets out a number of examples illustrating the intended operation of the recommended rules.

Notes

1. See the new Part 6A TIOPA 2010 (Taxation International and other Provisions) Act 2010, which came into effect on 1 January 2017 (the "UK Hybrids Rules") and the Council Directive amending Directive (EU) 2016/1164 as regards hybrid mismatches with third countries dated 12 May 2017 ("ATAD 2"), http://dsms.consilium.europa.eu/952/Actions/Newsletter.aspx?messageid=13108&customerid=37917&password=enc_643345636135526A32344361_enc (accessed on 13 June 2017).

2. See *The OECD releases a discussion draft on branch mismatch structures under Action 2 of the BEPS Action Plan* (22 August 2016): www.oecd.org/tax/aggressive/oecd-releases-discussion-draft-on-branch-mismatch-structures-under-action-2-of-the-beps-action-plan.htm (accessed on 31 May 2017).

Introduction

1. Branch mismatches arise where the ordinary rules for allocating income and expenditure between the branch and head office result in a portion of the net income of the taxpayer escaping the charge to taxation in both the branch and residence jurisdiction. Unlike hybrid mismatches, which result from conflicts in the legal treatment of entities or instruments, branch mismatches are the result of differences in the way the branch and head office account for a payment made by or to the branch. Because branch mismatches turn on differences in tax accounting rather than legal characterisation, the same basic legal structure may call for the application of different branch mismatch rules, depending on the accounting treatment adopted by the branch and head office.

2. This report identifies five basic types of branch mismatch arrangements:

    a. disregarded branch structures where the branch does not give rise to a permanent establishment (PE) or other taxable presence in the branch jurisdiction

    b. diverted branch payments where the branch jurisdiction recognises the existence of the branch but the payment made to the branch is treated by the branch jurisdiction as attributable to the head office, while the residence jurisdiction exempts the payment from taxation on the grounds that the payment was made to the branch

    c. deemed branch payments where the branch is treated as making a notional payment which results in a mismatch in tax outcomes under the laws of the residence and branch jurisdictions

    d. DD branch payments where the same item of expenditure gives rise to a deduction under the laws of both the residence and branch jurisdictions

    e. imported branch mismatches where the payee offsets the income from a deductible payment against a deduction arising under a branch mismatch arrangement.

Branch mismatches rules can arise directly as well as indirectly through a taxpayer's investment through a transparent structure such as a partnership.

Branch payee structures that give rise to D/NI outcomes

3. The first two categories of mismatches considered in this report are D/NI outcomes that arise where the residence jurisdiction treats a deductible payment as received through a foreign branch (and therefore excludes or exempts the payment from ordinary income) while the branch jurisdiction does not tax the payee because:

    a. in the case of a disregarded branch structure, the payee has an insufficient presence in the branch jurisdiction to be taxable on such payment

b. in the case of a diverted branch payment the branch jurisdiction exempts or excludes the payment from taxation on the grounds that the payment is treated as made to the head office.

Both these structures are discussed in further detail below.

Disregarded branch structure

4. In a disregarded branch structure the mismatch arises due to the fact that a deductible payment received by a taxpayer is treated, under the laws of the residence jurisdiction, as being made to a foreign branch (and therefore eligible for an exemption from income), while the branch jurisdiction does not recognise the existence of the branch and therefore does not subject the payment to tax. An example of a disregarded branch structure is illustrated in Figure 1.

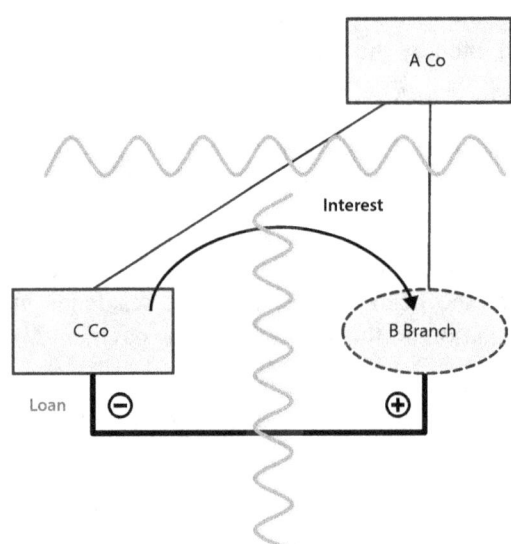

Figure 1. **Disregarded branch structure**

5. In this case A Co lends money to C Co (a related company) through a branch located in Country B. Country C permits C Co to claim a deduction for the interest payment. Country A exempts or excludes the interest payment from taxation on the grounds that it is attributable to a foreign branch. The interest income is not, however, taxed in Country B because A Co does not have a sufficient presence in Country B to be subject to tax in that jurisdiction. The payment of interest therefore gives rise to an intra-group mismatch (a D/NI outcome).

6. The D/NI mismatch that results from a disregarded branch structure can arise in a number of ways and could be a product of the domestic rules operating in each jurisdiction or due to a conflict between domestic law and treaty requirements. For example:

   a. The interest payment could be treated as income of a foreign branch (and therefore tax exempt) under Country A domestic law but may not be included in income under Country B domestic law because the branch does not give rise to taxable presence in Country B for domestic law purposes.

b. The branch could be treated as constituting a permanent establishment (PE) under the Country A-B tax treaty so that Country A is required to exempt the interest payment from tax under a provision equivalent to Article 23A of the OECD Model Tax Convention on Income and Capital: Condensed Version 2014[1] (Model Tax Convention, OECD 2014) (even though the branch does not give rise to a taxable presence under Country B's domestic law).

c. The branch may not meet the legal definition of a PE under the Country A-B tax treaty so that the payment of interest received by the branch is excluded from taxation by Country B because a provision equivalent to Article 7 of the Model Tax Convention (OECD, 2014) does not allow Country B to tax residents of Country A in the absence of a PE as defined under that treaty. This may be the outcome provided for under the treaty even though Country A's domestic law allows A Co to treat the payment as exempt from tax in Country A as income of a foreign branch.

7. The mechanics and the resulting tax outcomes from the use of a disregarded branch structure are similar to those of a reverse hybrid (discussed in Chapters 4 and 5 of the Action 2 Report (OECD, 2015)) in that both the residence and the branch jurisdiction exempt or exclude the payment from income on the grounds that the payment should be treated as received (and therefore properly subject to tax) in the other jurisdiction.

Diverted branch payment

8. A diverted branch payment has the same structure and outcomes as a payment to a disregarded branch except that the mismatch arises, not because of a conflict in the characterisation of the branch, but rather due to a difference between the laws of the residence and branch jurisdiction as to the attribution of payments to the branch. An example of a diverted branch payment is illustrated in Figure 2. This example is the same as that described in Figure 1, except that both the residence and branch jurisdiction recognise the existence of the branch. The mismatch arises from the fact that the branch treats the deductible interest payment as if it was paid directly to the head office in Country A, while the head office continues to treat the payment as made to the branch. As a consequence, the payment is not subject to tax in either jurisdiction (a D/NI outcome).

Figure 2. **Diverted branch payment**

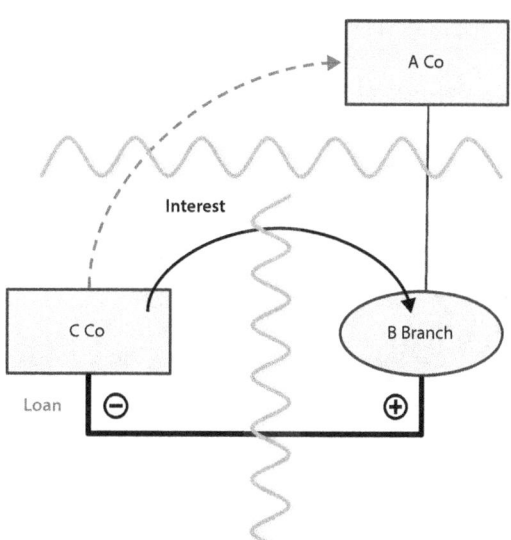

9. This mismatch in tax treatment could be due to a difference in the rules used by Country A and B for allocating income to the branch (or a difference in the interpretation or application of those rules) or due to specific rules in Country B that exclude or exempt this type of income from taxation at the branch level due to the fact that the payment is treated as made to a non-resident. As with the disregarded branch structures, the mechanism by which the mismatch in tax outcome arises is similar to that of a reverse hybrid in that both the residence and the branch jurisdiction exempt or exclude the payment from taxation on the basis that it should properly be regarded as received in the other jurisdiction.

Deemed branch payments

10. In the case of diverted or disregarded branch payments the mismatch arises in respect of a deductible payment that is not included in income in either the branch or residence jurisdiction. It is also possible, however, to generate internal mismatches between the branch and residence jurisdictions where the rules in those jurisdictions for allocating net income between the branch and head office permit the taxpayer to recognise a deemed payment between two parts of the same taxpayer and there is no corresponding adjustment to the net income in the payee jurisdiction that takes into account the effect of this payment.

11. A structure illustrating a deemed branch payment is set out in Figure 3. In this example A Co supplies services to an unrelated company (C Co) through a branch located in Country B. The services supplied by the branch exploit underlying intangibles owned by A Co. Country B attributes the ownership of those intangibles to the head office and treats the branch as making a corresponding arm's length payment to compensate A Co for the use of those intangibles. This deemed payment is deductible under Country B law but is not recognised under Country A law (because Country A attributes the ownership of the intangibles to the branch). Meanwhile, the services income received by the branch is exempt from taxation under Country A law due to an exemption or exclusion for branch income in Country A.

Figure 3. **Deemed branch payment**

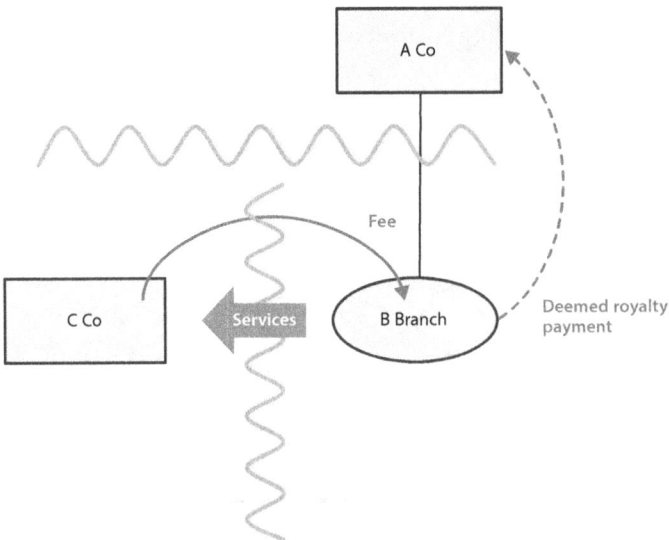

12. The deemed payment will give rise to an intra-group mismatch (a D/NI outcome) to the extent the deduction is set off against branch income which is exempt from tax in Country A (non-dual inclusion income). Deemed branch payments can only arise in those cases where the rules for allocating net income to the branch or head office allow for the recognition of notional payments between various parts of the same taxpayer. While the structure illustrated above involves a deemed royalty payment, the application of tax or accounting principles as well as income allocation principles in the branch jurisdiction can also give rise to other deemed payments (such as interest) with similar tax consequences.

13. The mismatches that arise in respect of deemed branch payments are similar to those that arise in respect of disregarded hybrid payments described in Chapter 3 of the Action 2 Report (OECD, 2015). In that case a hybrid payer (a person that is treated as a separate entity under the laws of the payer jurisdiction but as transparent or disregarded by the payee) makes a deductible payment that is disregarded under the laws of the payee jurisdiction due to the transparent tax treatment of the payer. The deduction resulting from that payment is then set off against income that is not subject to tax in the payee jurisdiction (i.e. against non-dual inclusion income).

14. The mechanics of, and outcomes resulting from, deemed branch and disregarded hybrid payments are substantially the same. The branch is entitled to a deduction for an item that is treated as expenditure under the laws of the payer/branch jurisdiction but that is disregarded in the payee/residence jurisdiction because the payee does not treat the payer as a separate enterprise for tax purposes. The deduction that is attributable to the mismatch is then set off against non-dual inclusion income, giving rise to a mismatch in tax outcomes.

DD branch payments

15. DD outcomes arise where the same item of expenditure is treated as deductible under the laws of more than one jurisdiction. These type of mismatches give rise to tax policy concerns where the laws of both jurisdictions permit the deduction to be offset against income that is not taxable under the laws of the other jurisdiction (i.e. against non-dual inclusion income).

16. DD branch payments can arise where the residence jurisdiction provides the head office an exemption for branch income while permitting it to deduct the expenditures attributable to the branch. Mismatches can arise where the rules for allocating income and expenditure in the branch jurisdiction also allow the taxpayer to claim a deduction for the same expenditure under the laws of the branch jurisdiction. In these cases the general exemption for branch profits provided by the residence jurisdiction means that the deduction in the branch will be set off against income that is not subject to tax in the residence jurisdiction (i.e. against non-dual inclusion income).

17. DD branch payments can also arise in the context of taxable branches (i.e. where the residence jurisdiction brings all the income and expenditure of the branch into account for tax purposes). Taxable branches can be used to generate DD branch outcomes where the branch is permitted to join a tax group or there is some other mechanism in place in the branch jurisdiction that allows expenditure or loss to be set off against income derived by another person that is not taxable under the laws of the residence jurisdiction.

18. In the example illustrated in Figure 4, A Co has established both a branch operation and a subsidiary in Country B. Country B law permits the subsidiary (B Co) and the Country B Branch to form a group for tax purposes, which allows the expenditure incurred by the Country B Branch to be offset against the income of the subsidiary.

Figure 4. **DD branch payment**

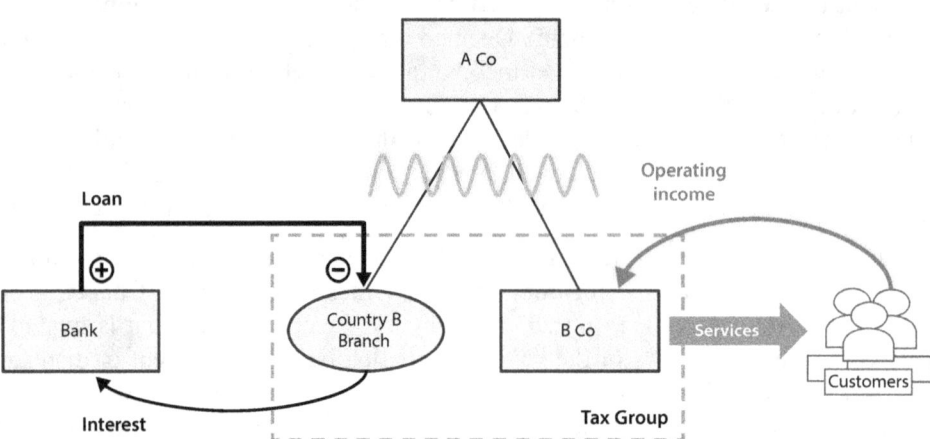

19. If Country B Branch is treated as taxable under the laws of Country A, then the interest expense incurred by the branch will give rise to separate deductions under the laws of Country A and Country B. Because Country B Branch and B Co are members of the same tax group this interest expenditure can also be offset, under Country B law, against the operating income derived by the subsidiary (i.e. against non-dual inclusion income). This structure therefore permits the same interest expense to be set off simultaneously against different items of income in the residence and branch jurisdiction.

20. The issues raised by these structures are discussed in Chapter 6 of the Action 2 Report (OECD, 2015) which sets out general hybrid mismatch rules neutralising the effect of DD outcomes. While the recommendations set out in Chapter 6 are drafted broadly enough to cover DD outcomes arising in respect of branch structures, the Action 2 Report (OECD, 2015) does not specifically consider the application of the deductible hybrid payments rule to DD branch payments such as those identified above.

Imported branch mismatches

21. An imported branch mismatch can arise where a person with a deduction under a branch mismatch arrangement offsets that deduction against a taxable payment received from a third party. An example of an imported branch mismatch is illustrated in Figure 5. This example is similar to that illustrated in Figure 3 except that A Co and C Co are part of the same group and it is assumed that there is no rule in either Country A or B addressing the mismatch in tax outcomes arising from a deemed royalty payment. As a consequence, a deduction under a branch mismatch arrangement is set off against the (deductible) service fee paid by C Co resulting in an indirect D/NI outcome.

22. The structure is similar to the imported mismatch structures described in Recommendation 8 of the Action 2 Report (OECD, 2015) in that it relies on the taxpayer engineering a mismatch (in this case a branch mismatch) under the laws of two jurisdictions and importing the effect of that mismatch into a third jurisdiction through a plain-vanilla instrument with an otherwise orthodox tax treatment.

23. Imported branch mismatch structures raise similar tax policy issues to those identified in the Action 2 Report (OECD, 2015) in that the most appropriate and effective way to neutralise the mismatch is for either or both Country A and B to implement branch

mismatch rules neutralising the mismatch. However, in order to maintain the integrity of the other recommendations (in the event Country A or B do not have branch mismatch rules), an imported mismatch rule is needed to deny the deduction for any payment that is directly or indirectly set off against any type of branch mismatch payment.

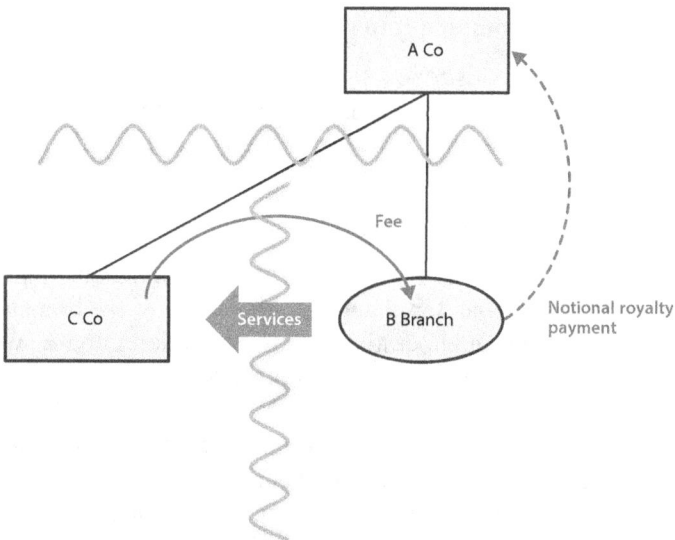

Figure 5. **Imported branch mismatches**

Summary of Recommendations

24. This report is divided into five chapters that set out specific recommendations for improvements to domestic law designed to reduce the frequency of branch mismatches as well as targeted branch mismatch rules, which neutralise the mismatch in tax outcomes without disturbing any of the other tax, commercial or regulatory outcomes. The recommendations set out in each chapter are summarised briefly below:

  a. Chapter 1 contains specific recommendations regarding the scope and operation of the branch exemption intended to achieve a closer alignment between that exemption and the policy of exempting the income of a foreign branch as a method of relieving income from double taxation (**Recommendation 1**).

  b. Chapter 2 sets out the operation of the branch payee mismatch rule which denies the payer a deduction for a diverted or disregarded branch payment made to a related person or under a structured arrangement to the extent the payment is not included in income by the payee (a rule that is equivalent to the reverse hybrid rule set out in Chapter 4 of the Action 2 Report (OECD, 2015)) (**Recommendation 2**).

  c. Chapter 3 describes the deemed branch payment rule which denies a deduction for a deemed payment between the branch and the head office (or between two branches of the same person) to the extent that payment gives rise to a D/NI outcome and the resulting deduction is set off against non-dual inclusion income (a rule that is equivalent to the disregarded hybrid payment rule set out in Chapter 3 of the Action 2 Report (OECD, 2015)) (**Recommendation 3**).

d. Chapter 4 clarifies the scope of the double deduction rule set out in Chapter 6 of the Action 2 Report (OECD, 2015) in respect of DD outcomes arising from payments made by a branch (**Recommendation 4**).

e. Chapter 5 provides for an imported mismatch rule consistent with Recommendation 8 in the Action 2 Report (OECD, 2015) that would deny a deduction for a payment made within the same control group or under a structured arrangement to the extent the income from such payment is set off against expenditure giving rise to a branch mismatch (**Recommendation 5**).

25. The recommendations in this report follow the same structure of those set out in the Action 2 Report (OECD, 2015) and, accordingly, any technical terms that are not defined in this report have the same meaning as the terms used in Action 2 Report (OECD, 2015).

Recommendation 1 not a branch mismatch rule

26. The recommendations described in Chapter 1 are not branch mismatch rules. Rather they are specific recommendations for changes to the scope of the branch exemption that are designed to bring the scope and operation of that exemption into line with the intended policy of avoiding double taxation of branch income. While narrowing the scope of the branch exemption will have the effect of reducing the frequency of branch mismatches (and therefore the need to apply any of the recommended branch mismatch rules set out in Chapters 2 to 5 of the report), the recommendations in Chapter 1 do not specifically target branch mismatches and apply to a wider range of payments than those targeted by the branch mismatch rules. The recommendations in Chapter 1 should not, however, be interpreted as requiring countries to make any change to deliberate policy decisions they have made, including in respect of the territorial scope of their tax regime, and do not purport to affect a country's obligations under a tax treaty.

Recommendations in Chapters 2 to 5 only require adjustments in respect of branch mismatches

27. The branch mismatch rules described in Chapters 2 to 5 are intended to neutralise mismatches that result from differences in the allocation of income or expenditure between the branch and the head office (or two parts of the same taxpayer). The rules should not apply when the reason for the mismatch is that the payee is exempt from tax, subject to a special tax regime or resident in a zero tax jurisdiction. Mismatches that arise solely due to differences in measurement or timing are also not within the intended scope of the recommendations.

Branch mismatch rules only to be applied after ordinary rules for allocating net income to the branch

28. Adjustments under the branch mismatch rules should only be made after applying the ordinary domestic rules for allocating branch income, subject to the requirements of any relevant treaty, but including any rules that restrict the scope of the branch exemption in accordance with the specific recommendations set out in Chapter 1. As branch mismatches are the result of taxpayers taking inconsistent positions in two jurisdictions on the same item of income or expenditure, there should generally be no need to apply branch mismatch rules where the taxpayer has adopted consistent positions and consistently applied the same standards to the allocation of branch income in both jurisdictions. The branch mismatch rules are intended to remove any incentive for a taxpayer to take inconsistent positions in

respect of where a payment is included or where functions are performed, assets are held and risks are assumed. The rules also eliminate the possibility of a taxpayer offsetting a deduction for the same expenditure against different items of income in two different jurisdictions. By neutralising these tax advantages, it is expected that taxpayers will adopt more consistent and coherent positions on the allocation of income and expenditure between the branch and the head office such that there will be little need to make many adjustments under these rules. Any adjustments under the recommendations set out in this report should not affect the allocation of taxing rights under a tax treaty.

29. The branch mismatch rules set out in this report introduce additional steps into the process of calculating the profit of the branch. This incremental compliance burden is likely to have a greater impact on substantial branches with commercial operations where there a large number of transactions in the branch with related and unrelated parties. Any such burden can, however, be minimised by taxpayers taking consistent positions on the allocation of income and expenditure between various parts of an enterprise and by jurisdictions ensuring that their existing domestic rules for allocating income and expenditure to a branch are clear, consistent and minimise the potential for both double taxation and double non-taxation. In the event that mismatches do arise, tax administrations should provide taxpayers with flexible and straight-forward implementation solutions that preserve the policy objectives behind the branch mismatch rules and that are based, as far as possible, on the taxpayer's existing domestic compliance and filing requirements. As with hybrid mismatch arrangements, the implementation solutions adopted by each jurisdiction should allow for effective and efficient co-ordination in the application of the branch mismatch rules in each jurisdiction without creating material gaps or the risk of double taxation.

30. Branch mismatches most frequently arise in the context of exempt branches (i.e. where the residence jurisdiction provides an exemption for branch income). Where a jurisdiction taxes residents on their worldwide income (including the income of any foreign branch), then any payments that are not included in income by the branch will generally be brought into charge in the residence jurisdiction (eliminating the risk of any branch payee mismatches). Furthermore, in the case of operating branches, the taxpayer will generally have sufficient dual inclusion income in the branch to avoid the need to make adjustments under the deemed branch payment or the double deduction rules described in Chapters 3 and 4 of this report (although there may still be scope for the operation of these rules where the branch jurisdiction permits the branch to join a tax group or provides some other mechanism, which allows the branch expenditure to be set off against non-dual inclusion income).

31. Branch mismatches can arise directly, where the same entity or person has taxable operations in a number of different jurisdictions, or indirectly through a taxpayer's participation through a transparent structure such as a partnership. Branch mismatch rules apply to a taxpayer in both these cases to neutralise the mismatch in tax outcomes.

Note

1. OECD Model Tax Convention on Income and Capital: Condensed Version 2014, OECD Publishing, Paris (Model Tax Convention, OECD 2014).

Bibliography

OECD (2015), *Neutralising the Effects of Hybrid Mismatch Arrangements, Action 2 – 2015 Final Report*, OECD Publishing, Paris, http://dx.doi.org/10.1787/9789264241138-en.

OECD (2014), *Model Tax Convention on Income and on Capital: Condensed Version 2014*, OECD Publishing, Paris, http://dx.doi.org/10.1787/mtc_cond-2014-en.

Chapter 1

Limitation to the scope of the branch exemption

> **1. Limitation to the scope of the branch exemption**
>
> Jurisdictions that provide an exemption for branch income should consider limiting the scope and operation of this exemption so that the effect of deemed payments, or payments that are disregarded, excluded or exempt from taxation under the laws of the branch jurisdiction, are properly taken into account under the laws of the residence jurisdiction.

Overview

32. Branch payee and deemed branch payment mismatches most frequently arise where the net income of the branch is exempt from tax in the residence jurisdiction. These risks can be significantly reduced if the residence jurisdiction modifies the operation of its branch exemption so as to ensure that the net income eligible for exemption is not greater than the income actually included by the branch. This can be done by including any items of income that are not taxed by the branch jurisdiction and by making the necessary adjustments to take into account the effect of deemed payments made from the branch to the head office. Changes to the scope of a branch exemption that required the taxpayer to make an adjustment in the residence jurisdiction in respect of a deemed payment or an item of income that was not taxable at the branch level, would provide for a comprehensive and transparent way of addressing branch mismatches and alleviate the payer from any need to consider whether an adjustment was required under the branch mismatch rules. This report therefore recommends that jurisdictions consider modifying the scope and operation of their branch exemption regime in order to take into account payments that are not included in income by an exempt branch and deemed payments made by an exempt branch.

33. There are a number of advantages to bringing a branch payment or deemed branch payment into income in the residence jurisdiction rather than relying on the rules set out in Chapters 2 to 5 of the report to address any mismatch in tax outcomes. From a compliance perspective, it will usually be easier for the head office to identify the payment or deemed payment that gives rise to the mismatch than it will be for the payer jurisdiction to apply the branch payee mismatch rule under Recommendation 2 or imported mismatch rule under Recommendation 5. Changes to the scope of the branch exemption also have the potential to eliminate a wider range of mismatches, including D/NI payments received from outside the controlled group and mismatches that result from the fact that the branch is exempt from tax, subject to a special regime or located in a jurisdiction that does not impose an income tax.

34. Some of the advantages of applying Recommendation 1 are discussed in **Example 1** of this report. In that example, a group company makes a deductible payment to the branch of another group company. The example notes there may be a number of reasons why the payment is not subjected to tax in the branch jurisdiction (e.g. the branch jurisdiction may not impose a corporate income tax, the payment may qualify for special treatment under a tax regime or the foreign branch may treat the payment as being made to the head office). Recommendation 1 could be applied to neutralise any resulting mismatch in all these cases. **Example 4** describes a deemed branch payment where the branch is allowed a deduction for a notional royalty payment made to the head office. The example notes that are a variety of methods that the residence jurisdiction could adopt to eliminate the risk of mismatches arising in respect of such notional payments that may be less complicated than applying the deemed branch payment rule.

35. Recommendation 1.1 is based on the assumption that the intention of the residence jurisdiction in granting an exemption for branch income is to relieve that income from double taxation, so that income that is not, in fact, subject to net taxation in the branch jurisdiction should not benefit from this exemption. Recommendation 1 should not, however, be interpreted as requiring countries to make any change to deliberate policy decisions they have made, including in respect of the territorial scope of their tax regime. Accordingly, this recommendation only calls for jurisdictions to consider modifying the scope and operation of their branch exemption to neutralise branch mismatches and does not set out any limitations on the amount of the adjustment, or the mechanism for making

that adjustment, provided any adjustment is consistent with a jurisdiction's tax treaty obligations, and tax policy settings in that jurisdiction.

Recommendation 1.1 – Limitation to the scope of the branch exemption

36. Recommendation 1.1 suggests jurisdictions consider narrowing the scope and adjusting the operation of their branch exemption regime in order to reduce the frequency of branch payee and deemed branch payment mismatches. The recommendation encourages the residence jurisdiction to consider limiting the operation and scope of the branch exemption so that the effect of any deemed payment or any payment that is not included in income under the laws of the branch jurisdiction is properly taken into account for tax purposes by making appropriate adjustments in the residence jurisdiction. As with Recommendation 5.1 of the Action 2 Report (OECD, 2015), this recommendation is designed to ensure that the branch exemption operates in line with the intended tax policy settings in the residence jurisdiction in respect of the taxation of worldwide income, while preserving the ability of jurisdictions to determine the scope of their taxing jurisdiction consistent with their general system of taxation.

37. While the purpose and effect of Recommendation 1.1 is to reduce the frequency of branch mismatches, this recommendation is not a branch mismatch rule. Rules that adjust the scope of the branch exemption in order to reduce instances of double non-taxation could apply to any payment that would ordinarily give rise to income in the residence jurisdiction, regardless of whether that payment produces a mismatch in tax outcomes or whether the mismatch in question is attributable to differences in the rules for allocating such payments between the branch and the head office. This is illustrated in **Example 1** where it is noted that the residence jurisdiction may choose to bring untaxed branch income into the charge to tax not only in those cases where the reason for mismatch is due to a misallocation of the payment under the laws of the branch jurisdiction, but also where the payment qualifies for tax-free treatment in the branch on some other basis.

38. Requiring the taxpayer to bring make an adjustment in the residence jurisdiction that takes into account the effect of the deemed or untaxed payment will not automatically trigger an additional tax liability in that jurisdiction. For example, under this rule a payment, such as a dividend, that was not taxed at the branch level (and was therefore required to be brought into account for tax purposes by the head office) may still be eligible to benefit from a tax exemption or other type of tax relief in the residence jurisdiction that is provided for payments of that nature under domestic law (such as a participation exemption for foreign dividends).

39. As with Recommendation 5.1 there are a number of ways the residence jurisdiction could make an adjustment to include an appropriate amount of additional income under the laws of the residence jurisdiction in order to neutralise any double non-taxation outcome and accordingly Recommendation 1.1 does not extend to describing the way in which the payment of untaxed branch income may be taken into account in the head office. **Example 1** considers the case of a licence fee paid to another group company that is not brought into tax in either the branch or the residence jurisdiction. The example notes that there are a variety of adjustments the residence jurisdiction could take to expand the scope of its taxing regime to bring untaxed branch income into charge at the head office. **Example 4** describes a deemed branch payment where the branch jurisdiction allows the branch a deduction for a notional royalty payment made to the head office. That example notes that there are a variety of methods for allocating income and expenditure between the head office and branch that can be used in order to take into account the effect of

such a deemed payment. These include recognising additional income in the head office jurisdiction of an amount equal to the deemed payment, allocating expenditure of an equivalent category to the payer jurisdiction and adjusting the way in which exempt income of the branch is calculated so as to eliminate the risk of mismatches arising in respect of such notional payments. In all cases, the adjustments required by the residence jurisdiction should be consistent with a proper allocation of income and expenditure between the branch and the head office under agreed international standards and in line with the intended territorial scope of that jurisdiction's tax regime.

40. It should also be noted that the residence jurisdiction may be prevented from restricting the scope of the branch exemption in those cases where the tax treaty in effect between the residence and branch jurisdiction contains a provision equivalent to Article 23A of the Model Tax Convention (Model Tax Convention, OECD 2014).

Bibliography

OECD (2015), *Neutralising the Effects of Hybrid Mismatch Arrangements, Action 2 – 2015 Final Report*, OECD Publishing, Paris, http://dx.doi.org/10.1787/9789264241138-en.

OECD (2014), *Model Tax Convention on Income and on Capital: Condensed Version 2014*, OECD Publishing, Paris, http://dx.doi.org/10.1787/mtc_cond-2014-en.

Chapter 2

Branch payee mismatch rule

1. Denial of deduction for branch payee mismatches

The payer jurisdiction should deny a deduction for a payment that gives rise to a D/NI outcome to the extent that the mismatch is a result of:

   a. differences in the allocation of payments between the residence and the branch jurisdiction or between two branch jurisdictions; or

   b. the fact that the payment is to a disregarded branch.

2. Disregarded branch

A disregarded branch is a branch that is treated as giving rise to a taxable presence under the laws of the residence jurisdiction (and thus is eligible for an exemption from income) but is not treated as giving rise to a taxable presence under the laws of the branch jurisdiction.

3. Scope

This recommendation shall only apply to payments made under a structured arrangement or between members of a controlled group.

Overview

41. A deductible payment made to a branch will give rise to a D/NI outcome where that payment is not included in ordinary income by either the residence or branch jurisdiction. The branch payee mismatch rule neutralises these types of mismatches where they result from both jurisdictions treating the payment as allocated to a taxpayer in the other jurisdiction.

42. Recommendation 2 specifically targets the two types of branch payee mismatches identified in the Introduction:

   a. **Diverted branch payments**, where the mismatch arises, not because of a conflict in the characterisation of the branch, but rather, due to difference between the laws of two jurisdictions as to the attribution of payments to the branch.

   b. **Disregarded branch structures**, where the mismatch arises due to the fact that a deductible payment received by a taxpayer is treated, under the laws of the residence jurisdiction, as being made to a foreign branch (and therefore eligible for an exemption from income) while the branch jurisdiction does not recognise the existence of the branch and therefore does not subject the payment to tax.

43. The mechanics and the resulting tax outcomes from the use of a disregarded branch structure and diverted branch payments are similar to the use of a reverse hybrid (discussed in Chapter 4 of the Action 2 Report (OECD, 2015)) in that both of the payee jurisdictions exempt or exclude a payment from income on the grounds that the payment should be treated as received (and therefore properly subject to tax) in the other jurisdiction. The branch payee mismatch rule set out in this chapter brings the treatment of diverted branch payments and disregarded branch structures into line with the outcomes provided for under the reverse hybrid rule by denying the deduction for such payments to the extent the allocation of payments between the two jurisdictions gives rise to a mismatch in tax outcomes.

Recommendation 2.1 – Denial of deduction for branch payee mismatches

Payment

44. The definition of payment set out in Recommendation 2.1 of this report is intended to have the same meaning as that set out in the Action 2 Report (OECD, 2015). It includes a broad range of current expenditures such as rents, royalties, interest, payments for services and other payments that may be set off against ordinary income under the laws of the payer jurisdiction. The term would not typically cover the cost of acquiring an asset and would not extend to an allowance for a depreciation or amortisation.

D/NI outcome

Branch payee mismatch rule applies in any jurisdiction where payment is deductible

45. The definition of deduction set out in Recommendation 2.1 of this report is intended to have the same meaning as that set out in the Action 2 Report (OECD, 2015). A payment is deductible to the extent a jurisdiction allows the taxpayer to offset expenditure against a taxpayer's ordinary income. The definition in the Action 2 Report (OECD, 2015) focuses on whether a payment falls into the category of a "deductible" item under the laws of the

relevant jurisdiction so that the specific details of the taxpayer's net income calculation should not generally affect the question of whether a payment is treated as "deductible" for tax purposes.

46. A payment may be treated as made from more than one jurisdiction in those cases where the payment is made through a tax transparent structure such as a branch or hybrid entity. In these cases the question of whether the payment gives rise to a D/NI outcome under the laws of the jurisdiction applying the branch payee mismatch rule is not affected by fact that the payment may also be deductible under the laws of another jurisdiction. This principle is the same as that illustrated in Example 4.4 of the Action 2 Report (OECD, 2015) where a hybrid entity makes a payment to a reverse hybrid. In this case the example concludes that the hybrid mismatch rule in Recommendation 4 of the Action 2 Report (OECD, 2015) should be applied in both the parent and subsidiary jurisdictions to neutralise the effect of the mismatch.

Not included in income in the head office or any branch

47. While the branch payee mismatch rule is the primary (and, in effect, only) branch mismatch rule for neutralising payments to a branch payee, this rule will not be triggered in the payer jurisdiction unless the payment actually gives rise to a D/NI outcome. As with the reverse hybrid rule described in Chapter 4 of the Action 2 Report (OECD, 2015), if the payment is brought into account as ordinary income in at least one jurisdiction then there will be no mismatch for the rule to apply to. This will be the case where the mismatch has been neutralised by a rule in the branch or head office jurisdiction which ensures that the payment that is not brought into account in one jurisdiction must be brought into account in the other. This would include any rule, consistent with Recommendation 1.1 of this report, that restricted the scope of branch exemption in the residence jurisdiction to payments that had actually been brought into the charge to taxation by the branch. **Example 1** considers the case of a licence fee that is paid to a branch of a company within the same control group as the payer. The example notes that the branch payee mismatch rule should not apply where the mismatch has been neutralised by a rule in the residence jurisdiction which ensures that any payment that is not brought into account in the branch must be brought into account in the head office. Thus if the residence jurisdiction, in accordance with Recommendation 1.1, restricts the scope of a branch exemption to payments that have actually been brought into the charge to taxation by the branch then the mismatch in tax outcomes would be neutralised and there will be no scope for the operation of the branch payee mismatch rule.

48. It should be borne in mind, when applying the branch payee mismatch rule, that the rule is not intended to address mere differences in timing, so that a deduction claimed for a payment in one taxable period should not be treated as giving rise to a mismatch simply because the payment will not be included until a subsequent period. It will be the payer who has the burden of establishing, to the reasonable satisfaction of the tax administration, that the rules of the payee jurisdiction require the payment to be brought into income, although it is expected that the tax position of the payee would usually be confirmed by means of a contractual representation.

49. The test for whether a payment has been "included" for tax purposes should be the same as that described in the Action 2 Report (OECD, 2015). A payment will be treated as included in the branch or head office (and therefore outside the scope of the branch payee mismatch rule) if, after a proper determination of the character and treatment of the payment under the laws of the relevant jurisdiction, the payment can properly be

considered to have been incorporated into a calculation of the payee's ordinary income. A payment that is taken into account by the payee under general law should not be treated as included if it benefits from a specific exclusion or exemption from tax on the grounds that the payment was made to a non-resident or a foreign branch.

50. In respect of commercial branch operations of a significant size, the volume of transactions and the complexity of the rules governing the calculation and allocation of income between the head office and branches may make it difficult for the taxpayer to establish to the satisfaction of a tax authority that a payment that has not been included in one jurisdiction, has been included in another. In these cases tax authorities may need to identify implementation solutions that are based, as much as possible, on existing domestic rules, administrative guidance, presumptions and tax calculations while still meeting the basic policy objectives of Recommendation 2. For example, a taxpayer may be able to demonstrate that the aggregate amount of income included for tax purposes in the head office and branch jurisdiction matches the (tax adjusted) income recognised in the accounts of the payee such that the tax authority is satisfied that all taxable payments made to such taxpayer have been recognised in at least one jurisdiction.

Inclusion under CFC or equivalent regime

51. The branch payee mismatch rule is only intended to operate where differences in the rules allocating payments between the branch and head office (or between two branches of the same person) give rise to a mismatch in tax outcomes. In certain cases, a payment to a branch that is not included by either the branch or head office may be included in the income of a parent company under a controlled foreign company (CFC) regime. Jurisdictions should consider the risk of economic double taxation in these cases and the extent of the adjustment that should be required under the branch payee mismatch rule in light of the fact that the payment is included in ordinary income under the CFC regime of a third country.

52. In those cases where the payer jurisdiction permits the taxpayer to rely on a CFC inclusion to limit the denial of the deduction in the payer jurisdiction, this exclusion should be limited to those cases where the taxpayer can satisfy the tax administration that the payment has been fully included under the CFC laws of the parent jurisdiction and is subject to tax at the full rate. This will include demonstrating that the payment is of a type that is ordinarily required to be brought into account under the relevant CFC rules and that the payment does not benefit from any exclusion (such as an active income exception). The taxpayer should also demonstrate that the quantification and timing rules of the CFC regime have actually brought that payment into account as ordinary income on the shareholder's return and may be required to show that the inclusion does not carry an entitlement to any unrelated foreign tax credit or other relief or even that the amount included is not set off against a deduction under another branch or hybrid mismatch arrangement (i.e. it does not give rise to an imported mismatch).

53. The treatment of payments that are included under a CFC regime is considered in **Example 1** in respect of a branch payee mismatch. In that case, although the intra-group payment is not included by either the residence or the branch jurisdiction, the example notes that it may be included in the income of the ultimate parent under a CFC (or equivalent) regime. If the payer jurisdiction wishes to avoid the risk of economic double taxation from denying a deduction for a payment that is, in fact, subject to tax under the CFC rules of another country, then it should consider the extent of the adjustment required under the branch payee mismatch rule in light of such CFC inclusion. The payer would

need, however, to satisfy the tax administration that the parent was actually required to include the payment under the relevant CFC rules and the payer may also need to satisfy the tax administration that the amount included under the CFC regime does not carry an entitlement to any unrelated foreign tax credit or other relief.

Counterfactual test to determine whether the mismatch is a result of misallocation of payment

54. As is the case for the reverse hybrid rule, the branch payee mismatch rule should not apply unless the payment would have been included as ordinary income if it had been paid directly to the head office. Example 4.1 of the Action 2 Report (OECD, 2015) provides an illustration of this principle in respect of an interest payment to a reverse hybrid. The example concludes that the reverse hybrid rule will not apply in cases where the investor is a tax exempt entity that would not have been subject to tax even if the payment had been made directly to the investor. The analysis and the outcomes described in that example are the same in the context of a diverted branch payment or a payment to a disregarded branch where the taxpayer is tax exempt under the laws of the residence jurisdiction. The same principle is applied in **Example 1** of this report in respect of branch payee mismatch. That example notes that the question of whether the mismatch is a result of the misallocation of payments between the branch and the head office can be answered by posing a counterfactual test that asks what the tax treatment of the payment would have been if it had been made directly to the head office. On the facts of **Example 1** it is the operation of the branch exemption that shelters the relevant payment from taxation under the laws of the residence jurisdiction, so that Recommendation 2 applies to deny a deduction for the payment in the payer jurisdiction if the payment is not subject to tax in the branch jurisdiction.

55. As with the reverse hybrid rule, this branch mismatch rule should not be used to circumvent the operation of the hybrid financial instrument rule and this rule should continue to apply to the extent a direct payment would have been subject to adjustment under Recommendation 1 of the Action 2 Report (OECD, 2015).[1]

Recommendation 2.2 – Disregarded branch

56. As described in detail in the Introduction of this report a disregarded branch is a branch that is treated as giving rise to a taxable presence under the laws of the head office jurisdiction (and thus is eligible for an exemption from income) but is not treated as giving rise to a taxable presence under the laws of the branch jurisdiction. Disregarded branch structures could be considered to be a subset of diverted branch payments given that the mismatch arises in respect of differences in the allocation of payments between the branch and head office. The difference between diverted branch payments and disregarded branches is that, in the case of disregarded branch structures, not only is there no inclusion of any payment by the branch jurisdiction, but there is nothing in the branch jurisdiction to attribute any payment to.

57. The "laws" referred to in Recommendation 2.2 include both domestic and treaty law. Therefore disregarded branches may arise in a situation where there are differences between the definition of a branch for domestic law and treaty purposes so that the branch is treated as constituting a permanent establishment (PE) under the relevant tax treaty (with the consequence that the head office is required to exempt the payment from tax under a provision equivalent to Article 23A of the Model Tax Convention) while the activities of

the branch do not result in the taxpayer having any taxable presence under the domestic laws of the branch jurisdiction. In these cases the residence jurisdiction may be prevented from restricting the scope of the branch exemption under Recommendation 1 owing to the overriding effect of the tax treaty. Alternatively the branch may not meet the legal definition of a permanent establishment under the tax treaty so that the payment of interest received by the branch is excluded from taxation by the branch jurisdiction because a provision equivalent to Article 7 of the Model Tax Convention (OECD, 2014) does not allow the branch jurisdiction to tax the payment in the absence of a PE as defined under that treaty. This may be the outcome provided for under the laws of the branch jurisdiction despite the fact that the residence jurisdiction treats the payment as received by a foreign branch and as eligible for an exemption from taxation under the domestic rules of the residence jurisdiction.

Recommendation 2.3 – Scope of the rule

58. The branch payee mismatch rule should only apply to payments made under a structured arrangement or between members of the same control group. In order to ensure consistency, the tests for "structured arrangement" and "control group" should be the same as those set out in the Action 2 Report (OECD, 2015). This would mean that a taxpayer would not be required to make an adjustment under the branch payee mismatch rule unless the payment was made to a person within the same control group or the payer was a party to a structured arrangement that was designed to produce a branch mismatch. As stated in the Action 2 Report (OECD, 2015):

> A person will be a party to a structured arrangement when that person has a sufficient level of involvement in the arrangement to understand how it has been structured and what its tax effects might be. A taxpayer will not be treated as a party to a structured arrangement, however, where neither the taxpayer, nor any member of the same control group, was aware of the mismatch in tax outcomes or obtained any benefit from the mismatch.[2]

59. A taxpayer may enter into a number of on-market transactions with unrelated parties that give rise to D/NI outcomes and the payer may not have the capacity to undertake due diligence on the transaction to determine whether there is a mismatch (or the reason for it). On-market transactions between unrelated parties will not, however, generally fall within the scope of the branch payee mismatch rules as the payer would generally be expected to enter these transactions on arm's length terms and could not be expected to make enquires as to a counterparty's tax position in the context of these type of trades.

60. Example 4.1 of the Action 2 Report (OECD, 2015) provides an illustration of the application of the reverse hybrid rule to an interest payment made by an unrelated third party. In that case, the example notes that the use of a reverse hybrid as a special purpose lending vehicle (SPV) may indicate that the arrangement between the investor and SPV has been engineered to produce a mismatch in tax outcomes. In that example, however, the payer is not treated as a party to that structured arrangement because it pays a market rate of interest under the loan and would not have been expected, as part of its ordinary commercial due diligence, to take into consideration the tax position of the counterparty when making the decision to borrow money. The same analysis and outcomes that apply to the reverse hybrid structure described in Example 4.1 should also apply to a similar example involving a diverted branch payment or a payment to a disregarded branch.

Notes

1. See Action 2 Report (OECD 2015) paragraph 167 and paragraph 11 of Example 4.4.
2. See Action 2 Report (OECD 2015), paragraph 342.

Bibliography

OECD (2015), *Neutralising the Effects of Hybrid Mismatch Arrangements, Action 2 – 2015 Final Report*, OECD Publishing, Paris, http://dx.doi.org/10.1787/9789264241138-en.

OECD (2014), *Model Tax Convention on Income and on Capital: Condensed Version 2014*, OECD Publishing, Paris, http://dx.doi.org/10.1787/mtc_cond-2014-en.

Chapter 3

Deemed branch payment rule

> **1. Denial of deduction for deemed branch payments**
>
> The jurisdiction that recognises a deemed branch payment (payer jurisdiction) should deny a deduction for that payment to the extent it gives rise to a branch mismatch.
>
> **2. Deemed branch payments**
>
> A deemed branch payment is a deemed payment between the branch and the head office or between two branches of the same taxpayer that gives rise to a D/NI outcome as a result of the fact that such payment is disregarded under the laws of the jurisdiction that is treated as receiving the payment (the payee jurisdiction).
>
> **3. No branch mismatch to the extent set off against dual inclusion income**
>
> A deemed branch payment shall give rise to a branch mismatch only to the extent the payer jurisdiction allows the deduction to be set off against an amount that is not dual inclusion income.

Overview

61. As described in the Introduction, a deemed payment between the branch and the head office (or between two branches) will give rise to a D/NI outcome where that payment is disregarded by the payee. This type of mismatch can give rise to tax policy issues where the payer jurisdiction allows the resulting deduction to be set off against an item of income that is not included under the laws of the payee jurisdiction (i.e. against income that is not "dual inclusion income"). The deemed branch payment rule in Recommendation 3 only applies where the payer jurisdiction allows the taxpayer to recognise notional payments between various parts of the same taxpayer. The rule neutralises any potential branch mismatch arising in respect of such a deemed branch payment by restricting the payer's deduction to the amount of dual inclusion income.

62. The deemed branch payment rule is intended to bring the treatment of deemed branch payments into line with the rules that apply to disregarded payments made by a hybrid entity under Recommendation 3 of the Action 2 Report (OECD, 2015). Unlike disregarded hybrid payments, however, where the deduction is a consequence of an actual payment between separate entities and the mismatch results from differences in the legal treatment of the payer under the laws of the payer and payee jurisdictions, a deemed branch payment is a purely notional payment between two parts of the same taxpayer resulting in a mismatch in the allocation of expenditure between the payer and payee jurisdictions and, accordingly, the rule will only apply in those jurisdictions that recognise such notional payments.

63. The fact that deemed branch payment mismatches are the product of a conflict in the rules for allocating expenditure (rather than in the legal characterisation of the payer) leads to a number of differences in the way the deemed branch payment rules operate. In particular, it means that deemed branch payment mismatches can generally be avoided by the head office jurisdiction adopting rules, in line with Recommendation 1, that result in an overall allocation of net income to the head office that is consistent with recognising the effect of the deemed payment. It also means that there is little (if any) scope for the application of a secondary (forced inclusion) rule in the context of deemed branch payments (see the commentary to Recommendation 3.1 below). Furthermore, the fact that the mismatch results from the misallocation of expenditure means that such mismatches can be neutralised by the payee jurisdiction allocating expenditure of an equivalent category to the payer jurisdiction (see the commentary to Recommendation 3.2 below).

Recommendation 3.1 – Denial of deduction for deemed branch payments

Deemed branch payment rule does not apply to depreciation or allowances for corporate equity

64. The deemed branch payment rule applies to deductions that result from notional payments to another part of the same taxpayer. These notional payments are tax fictions, used for determining the income that is properly subject to tax in the payer jurisdiction. Like the disregarded hybrid payment rules in the Action 2 Report (OECD, 2015), the deemed branch payment rules are not intended to apply to deductions for depreciation or losses in the value of an asset or domestic concessions such as allowances for contributed equity. While such allowances may be structured as a deduction from corporate income tax and the amount of that deduction may be calculated by reference to a notional amount (such as a risk-free rate of return on investment), their purpose is not to arrive at an accurate determination of the income that is properly subject to tax in the payer jurisdiction but

rather to unilaterally lower the effective rate of tax in order to encourage equity investment in that jurisdiction by reducing the tax distortions associated with the use of debt rather than equity.

No secondary rule under Recommendation 3

65. While the deemed branch payment rule requires the payer jurisdiction to deny a deduction for a deemed payment to the extent it gives rise to a branch mismatch, there is no corresponding secondary rule requiring the deemed payment to be included in income in the payee jurisdiction, as this is already the outcome provided for under Recommendation 1. **Example 4** describes a case where the branch jurisdiction allows a deduction for a notional royalty payment made by the branch to the head office. The example notes there are a variety of measures that the residence jurisdiction could adopt under Recommendation 1 that will result in the effect of the deemed payment being taken into account under the laws of the residence jurisdiction. These include: recognising an additional amount of income in the head office jurisdiction equal to the deemed payment; allocating expenditure of an equivalent category to the payer jurisdiction and/or adjusting the calculation of the net income of the branch so as to eliminate the risk of mismatches arising in respect of notional payments. If the residence jurisdiction adopts one of these measures, then the branch mismatch will not arise and there will be no scope for the application of the deemed branch payments rule.

Recommendation 3.2 – Deemed branch payments

Deemed payment

66. A deemed payment is any notional payment that is not calculated by reference to an actual expenditure of the taxpayer.

Notional payment

67. A notional payment is a payment that is treated as made between the branch and head office (or two branches) of the same taxpayer as part of profit allocation mechanism intended to arrive at an accurate determination of the income that is properly subject to tax in the payer jurisdiction. The payer jurisdiction is treated as making a notional payment to a branch or head office in respect of functions performed, assets held or risks assumed in the payee jurisdiction. The terms under which a notional payment is made may be documented as if the arrangement was between separate entities and accounted for through the transfer of funds between jurisdictions, however these notional payments do not have any independent legal status beyond giving effect to a proper allocation of net income between the payer and payee jurisdiction for tax purposes.

Calculated by reference to actual expenditure of the taxpayer

68. A notional payment should not be treated as a deemed payment to the extent it represents or is calculated by reference to actual expenditure recognised in the accounts of the taxpayer. A payment that is treated (for tax purposes) as made between the branch and the head office but which, in practice, represents an underlying third party expense should be treated as an actual payment rather than a deemed payment and therefore as outside the scope of the deemed branch payment rule.

69. A notional payment that is not expressly calculated by reference to actual expenditure should be treated as an actual payment where that payment relates to specific functions performed, assets held or risks assumed by another part of the same taxpayer and there is itemised expenditure of the same type in the accounts of the taxpayer, in respect of the same functions, risks or assets, which can be directly attributed to that deemed payment. In this case, where the notional payment can be defined with sufficient precision such that the purpose of the payment can be traced to an item of expenditure recorded in the taxpayer's accounts, then the taxpayer may treat the notional payment as an actual payment of the underlying expenditure incurred by the payee.

70. The approach described in the paragraph above is illustrated in **Example 5**, where the taxpayer contracts for various services from third party service providers. Part of these services includes software licences and IT support services relating to software owned by the taxpayer. The branch makes a notional royalty payment to the head office in respect of the same software. In this case, the nature of that services expenditure is such that it can accurately and reliably be attributed directly to the deemed payment. On this basis the taxpayer treats a portion of the deemed royalty payment as an actual payment for services supplied by third parties. In **Example 8** the taxpayer uses its own equity and money borrowed from an unrelated bank to make loans to customers located in the residence and branch jurisdictions. The branch jurisdiction treats the interest paid on the loans as attributable to the branch and also allows the branch a deduction for a deemed interest payment to the head office. While this payment is treated by the branch as a notional payment, if, in practice, the payment is calculated by reference to a certain percentage of the taxpayer's external borrowing costs or there is itemised interest expenditure or borrowing costs in the tax accounts of the payee that can directly attributed to that deemed payment then the interest expense claimed under the laws of the branch jurisdiction should not be treated as a deemed payment for the purposes of the deemed branch payments rule.

71. The fact that this type of payment is not caught by the deemed branch payment rule does not necessarily mean that the branch mismatch rules will not apply to that payment. Such a payment can still be caught by the double deduction rules in Recommendation 4. In **Example 5**, the deemed royalty payment that is characterised as expenditure attributed to third party services is also deductible under the laws of the residence jurisdiction, which means that the deduction triggers an adjustment under the double deduction rule. As demonstrated by **Example 9**, the fact that a notional interest payment is treated as an actual financing cost under the branch mismatch rules may result in the same adjustment being made in the branch jurisdiction under the secondary rule in Recommendation 4.

Deemed payment must be "disregarded" in the payee jurisdiction

72. A deemed payment will not give rise to a mismatch unless it is "disregarded" under the laws of the payee jurisdiction. In the case of a deemed payment the payee jurisdiction is the jurisdiction where the deemed payment is received or is treated as received. The payee jurisdiction may recognise a deemed payment by including the amount of the deemed payment as income or by the residence jurisdiction allocating expenditure or loss of an equivalent category to the payer jurisdiction and therefore disallowing the expenditure to be taken into account in that jurisdiction.

Recognition of payment by allocating equivalent category of expenditure or loss

73. Jurisdictions that exempt foreign source income will usually have corresponding rules that limit the deductibility of a taxpayer's expenses that are incurred in deriving that income. Where the residence jurisdiction has domestic rules limiting the deductibility of expenditure that has been incurred in deriving branch income then the effect of this limitation should be taken into account in determining the extent to which a deemed payment has been disregarded under the deemed branch payment rule. A residence jurisdiction that does not include a deemed payment directly in income should be treated as having recognised that payment as a payee jurisdiction to the extent it denies the head office a deduction for an equivalent category of expenditure, on the grounds that such expenditure has been allocated to the payer jurisdiction, provided such expenditure is not already treated as deductible in the branch.

74. The rules limiting deductibility of expenditure or loss may be applied by the head office jurisdiction on a case by case basis to each item of expenditure or loss or they may be the result of an allocation of a general category of expenditure between the head office and its branches. This allocation may be in accordance with a statutory or administrative formula that takes into account such factors as: the nature of the expenditure or loss (including the terms under which that expenditure or loss is incurred); the nature and extent of the activities in the branches and head office and the balance of assets and income in each jurisdiction.

75. Unlike the tracing approach described above, which is used to determine whether a notional payment represents or is calculated by reference to actual expenditure of the taxpayer, the determination of whether a deemed payment belongs to an equivalent category as an item of expenditure or loss in the head office jurisdiction is a broader test that should be done on a like-kind basis. Provided the deemed payment and allocated expenditure pertain to the same general category of assets, functions or risks (i.e. a straightforward explanation can be given for the relationship between the deemed payment and the allocated expenditure or loss) then the two items should be treated as belonging to an equivalent category for the purposes of Recommendation 3.

76. The deemed payment does not need to be of the same specific type as the expenditure or loss allocated by the head office and does not need to be calculated on the same basis in order to belong to an equivalent category. A deemed payment should only, however, be treated as recognised by the allocation of an equivalent category of expenditure or loss to the extent of the amount actually allocated to the payer jurisdiction and that the expenditure or loss has been denied in the payee jurisdiction as a result of such allocation.

77. In **Example 4** the taxpayer provides computer services to foreign customers through an exempt branch located in that country. Under the laws of the branch jurisdiction, the branch is permitted a deduction for a notional royalty payment made to the head office. This payment is intended to reflect an arm's length compensation for intellectual property that is exploited by the branch in the course of providing services to branch customers. The residence jurisdiction would have ordinarily allowed the head office a deduction for research and development (R&D) costs in respect of intellectual property used by the branch. In this case, however, the deduction is denied on the grounds that the income of the branch is exempt from tax under the laws of the residence jurisdiction. In this case, the deemed payment and allocated R&D costs pertain to the same general category of assets (the intellectual property that is being exploited by the branch) and the basis on which the R&D costs have been denied in the payer jurisdiction indicates that there is straightforward connection between the deemed payment and the allocated expenditure

or loss. Accordingly these two items are treated as being in an equivalent category for the purposes of deemed branch payment rule notwithstanding that the deemed payment (a royalty) is not the same type of expenditure that is allocated by the head office (R&D costs) and has not been calculated on the same basis.

78. In **Example 8** the branch jurisdiction allows the branch a deduction for a deemed interest payment to the head office. At the same time, the rules in the residence jurisdiction require the head office to treat a portion of the taxpayer's interest expense as attributable to the branch (and that portion is therefore non-deductible under the laws of the residence jurisdiction). In this case, both the deemed payment and the allocation of interest expenditure relate to the same general category of financing costs and accordingly the two items should be treated as being in an equivalent category for the purposes of the deemed branch payment rule. The example notes that even if the allocated financing costs in the residence jurisdiction relate to swap, derivative or guarantee fees they should still be treated as expenditure of an equivalent category, despite the fact that they are of a different type and calculated on a different basis.

79. As the domestic rules limiting deductibility will not necessarily be designed to accurately apportion expenditure to other jurisdictions, the taxpayer should be permitted to use the formula that is used to restrict the deductibility of expenditure (with any necessary adjustments) to calculate the amount that can be treated as allocated to a branch jurisdiction. This could be done, for example, by determining what the limitation on deductibility would have been in the branch jurisdiction had those limitation rules applied in that jurisdiction. For example, the head office may be subject to restrictions on interest deductibility on the basis that a portion of the borrowed funds have been used to support the activities of exempt foreign branches. In such a case the taxpayer could be permitted to apply the same interest limitation formula (with necessary adjustments) to the branch on a standalone basis to determine the amount of interest deduction that has been allocated to that branch.

Mismatch must be as a result of the fact that payment is disregarded

80. The deemed branch payments rule only applies where the reason for the D/NI outcome is the fact that the payment has not been recognised in the payee jurisdiction. This means that the rule should not apply, for example, where the payee would have benefitted from an exemption or exclusion in respect of that payment under the laws of the payee jurisdiction. In the context of the branch payee mismatch rule, this report applies a counterfactual test, which looks to what the tax treatment of the misallocated payment would have been, had it been included by the head office. The same counterfactual test cannot be applied in the context of Recommendation 3, where the payment does not have any independent legal status. Nevertheless, in order to achieve a parity of outcomes with the branch payee and disregarded hybrid payments rules, an adjustment should only be made where the payee is a person that is subject to tax under the laws of the payee jurisdiction.

Recommendation 3.3 – Rule only applies to payments that result in a branch mismatch

81. A deemed branch payment will not be treated as giving rise to a mismatch in tax outcomes if the deduction resulting from that payment, does not exceed dual inclusion income. The identification of whether an item should be treated as dual inclusion income

is primarily a legal question that requires an analysis of the treatment of the income under the laws of both jurisdictions. An amount should be treated as dual inclusion income if it is included in income under the laws of both jurisdictions even if there are differences in the way those jurisdictions value that item or in the accounting period in which the income is derived. In most cases it will be relatively straightforward for the payer jurisdiction to identify the items of income that are subject to tax under the laws of the payer and payee jurisdictions.

82. The set off of a deemed branch payment against an item of dual inclusion income is illustrated in **Example 2**. In that example, a deemed payment is made to the head office by a taxable branch (i.e. a branch whose income is fully subject to tax under the laws of the residence jurisdiction). The example notes that, in this case, where the operating income of the branch is included as ordinary income in both jurisdictions, there is likely to be limited scope for the application of the deemed branch payment rule because the deemed payment will generally be offset against dual inclusion income. In **Example 3**, the taxpayer restructures its operations in the branch jurisdiction and establishes a reverse hybrid entity to provide certain services to former branch customers. Although the restructuring reduces the amount of dual inclusion income under the structure, there is still no requirement for the branch jurisdiction to deny a deduction for the deemed payment under Recommendation 3 as the total amount of dual inclusion income under the structure still exceeds the amount of the deemed payment.

Foreign tax credits

83. An item that is treated as taxable income of a taxable branch should continue to be treated as dual inclusion income even when the residence jurisdiction allows a foreign tax credit for tax paid at the level of the branch. As stated in the Action 2 Report (OECD, 2015), in respect of disregarded hybrid payments:

> "Double taxation relief, such as a domestic dividend exemption granted by the payer jurisdiction or a foreign tax credit granted by the payee jurisdiction should not prevent an item from being treated as dual inclusion income where the effect of such relief is simply to avoid subjecting the income to an additional layer of taxation in either jurisdiction."[1]

The report notes, however, that such double taxation relief may give rise to policy concerns where it has the effect of generating surplus relief that may be offset against non-dual inclusion income.

84. While the payment of tax in the branch may give rise to a claim for direct foreign tax credits under the laws of the residence jurisdiction, these credits should not give rise to policy issues provided the residence jurisdiction has rules that limit the amount of direct foreign tax credits by reference to the total amount of foreign income in the branch. Such rules will generally prevent any surplus credit being offset against unrelated non-dual inclusion income. Direct foreign tax credits can, however, give rise to such surplus tax relief where the branch has both dual inclusion and non-dual inclusion income and the deemed payment results in a different basis for calculating the income of the branch under the laws of payer and payee jurisdictions. In this case, the payee jurisdiction could consider adjusting the amount of foreign income taken into account in determining the taxpayer's eligibility for a foreign tax credit to reflect the deduction claimed under the disregarded payment. The limitation on foreign tax credits in the payee jurisdiction is discussed in **Example 3** where it is noted that the residence jurisdiction may seek to limit the amount of the direct foreign tax credit the head office can claim in respect of income

from a taxable branch to the (adjusted) net income of the branch after taking into account the effect of any notional payments that have not been recognised by the head office. In the absence of any such limitation in the residence jurisdiction, the branch jurisdiction may consider restricting the definition of dual inclusion income, so as not to include income that has been sheltered from tax in the residence jurisdiction by surplus foreign tax credits (i.e. tax credits on income that has not, in fact, been included under the laws of the branch jurisdiction). Countries that introduce rules limiting the availability of foreign tax credits or restricting the definition of dual inclusion income in these cases should seek to strike a balance between rules that minimise compliance costs, preserve the intended effect of such double taxation relief and prevent taxpayers from entering into structures that undermine the integrity of the branch mismatch rules. Recommendation 3 should not, however, be interpreted as requiring countries to make any change to deliberate policy decisions they have made regarding the territorial scope of their tax regime. Accordingly, this recommendation only calls for jurisdictions to consider modifying the scope of their foreign tax credit rules to eliminate branch mismatches so far as those changes are consistent with the other tax policy settings in that jurisdiction.

Note

1. See Action 2 Report (OECD, 2015), paragraph 126.

Bibliography

OECD (2015), *Neutralising the Effects of Hybrid Mismatch Arrangements, Action 2 – 2015 Final Report*, OECD Publishing, Paris, http://dx.doi.org/10.1787/9789264241138-en.

Chapter 4

Double Deduction Rule

> **1. Treatment of Double Deduction Outcomes**
>
> To the extent a double deduction outcome gives rise to a branch mismatch:
>
> a. the deduction should be denied in the investor jurisdiction; and
>
> b. where the deduction is not denied in the investor jurisdiction, then the deduction should be denied in the payer jurisdiction.
>
> Any deduction should, however, be eligible to be offset against dual inclusion income whether arising in a current or subsequent period.
>
> **2. Double Deduction Outcome**
>
> A double deduction outcome means a deduction of the same payment, expense or loss in both the jurisdiction where such payment is made, expense is incurred or loss is suffered (the payer jurisdiction) and another jurisdiction (the investor jurisdiction).
>
> **3. No branch mismatch to the extent set off against dual inclusion income**
>
> A double deduction will give rise to a branch mismatch only to the extent the payer jurisdiction allows the deduction to be set off against an amount that is not dual inclusion income.

Overview

85. A taxpayer which incurs expenditure under a cross-border structure (including through a foreign branch) may be entitled to deduct that expenditure under the laws of two or more jurisdictions. This double deduction (DD) outcome will give rise to tax policy concerns where the laws of both jurisdictions permit the deduction to be set off against an amount that is not treated as income under the laws of the other jurisdiction (i.e. against income that is not "dual inclusion income"). The policy of the double deduction rule is to limit a taxpayer's deduction to the amount of dual inclusion income in circumstances where the deduction that arises in the other jurisdiction is not subject to equivalent restrictions.

86. As noted in the Introduction, the issues raised by DD outcomes are addressed in Chapter 6 of the Action 2 Report (OECD, 2015) which sets out hybrid mismatch rules neutralising their effect. While the recommendations set out in Chapter 6 are drafted broadly enough to cover DD outcomes arising in respect of branch structures, the Action 2 Report (OECD, 2015) does not consider, in any detail, the application of the deductible hybrid payments rule to expenditure incurred through a branch.

87. Recommendation 4 of this report clarifies the intended scope of the deductible hybrid payments rule in the Action 2 Report (OECD, 2015) by restating and clarifying the operation of that rule in the context of branch structures. This recommendation supplements, and does not replace, Chapter 6 of the Action 2 Report (OECD, 2015) and uses language that is consistent with ATAD 2[1]. In most cases, it is expected that countries would address DD outcomes involving the use of hybrid entities and branches under the same rules.

Recommendation 4.1 – Treatment of DD outcomes

88. The primary recommendation under the double deduction rule is that the investor (i.e. residence) jurisdiction should restrict the deductibility of any payment, expense or loss that is also deductible under the laws of the payer (i.e. branch) jurisdiction so that such amount can only be set off against dual inclusion income. The defensive rule, which imposes the same type of restriction in the payer jurisdiction, will only apply in the event that the effect of the mismatch is not neutralised in the investor jurisdiction. These rules apply when there is a branch under the laws of the payer jurisdiction regardless of whether the residence jurisdiction also recognises a branch in the payer jurisdiction.

89. Recommendation 4.1 allows excess deductions that are subject to restriction under the double deduction rule to be carried-forward to another period, in accordance with a jurisdiction's ordinary rules for the treatment of net losses, and applied against dual inclusion income in that period. This mirrors Recommendation 6 in Action 2 Report (OECD, 2015) for the deductible hybrid payments rule. Because the rule only applies to double deductions to "the extent the payer jurisdiction allows the deduction to be set off against" non-dual inclusion income, the rule does not limit the deductibility of stranded losses (see discussion under Recommendation 4.3 below).

Recommendation 4.2 – DD outcome

90. Unlike Recommendations 2 and 3, which apply to payments or deemed payments that give rise to D/NI outcomes, double deductions can also arise in respect of non-cash items such as depreciation or amortisation.

91. The double deduction rule should only operate to the extent a taxpayer is actually entitled to a deduction for a payment under local law. Accordingly the rule will not apply to the extent the taxpayer is subject to transaction or entity specific rules in the investor or payer jurisdictions that prevents the payment from being deducted. These restrictions on deductibility may include hybrid or branch mismatch rules that deny the taxpayer a deduction in order to neutralise a direct or indirect D/NI outcome.

92. If a payment has triggered a deduction under the laws of two or more jurisdictions, then differences between the rules used in the payer and investor jurisdictions for determining the value of that payment will not generally impact on the extent to which a payment has given rise to a mismatch in tax outcomes. Similarly the operation of the double deduction rule is not dependent on the timing of the deduction or receipt in the other jurisdiction.

93. Determining which payments have given rise to a double deduction (and which items are dual-inclusion income) requires a comparison between the domestic tax treatment of these items and their treatment under the laws of the other jurisdiction. It may be possible to undertake a line-by-line comparison of each item of income or expense in straightforward cases where the branch is performing limited functions (see **Example 9**). In more complex cases, however, where the taxpayer has entered into a number of transactions through the branch that give rise to different types of income and expense, countries may wish to adopt a simpler implementation solution for tracking double deductions and dual inclusion income. The way in which double deduction outcomes will arise will differ from one jurisdiction to the next and countries should choose an implementation solution that is based, as much as possible, on existing domestic rules, administrative guidance, presumptions and tax calculations while still meeting the basic policy objectives of Recommendation 4.

94. In the case of commercial branch operations it will generally be impractical for a taxpayer to adopt a line-by-line comparison of income and expenditure to determine whether the amount of double deductions exceeds the amount of dual inclusion income. In this case, the taxpayer could determine the amount of double deductions on an aggregate basis by comparing the total deductions claimed for actual expenditure and loss in each jurisdiction against the taxpayer's total relevant expenditures. This excess may be treated as a double deduction (subject to adjustment under Recommendation 4) to the extent it cannot be explained by reference to differences in timing or valuation. This comparison could be done on a category by category basis, a branch by branch or a whole of entity basis, however, the taxpayer should only be expected to make the adjustment in one jurisdiction.

95. **Example 6** and **Example 7** both illustrate the application of the double deduction rule to an entity with operating branches. These branches incur expenditure which gives rise to excess deductions. In both cases, the relevant excess is treated as a double deduction (subject to adjustment under Recommendation 4) to the extent it cannot be explained solely by reference to differences in timing or valuation. In **Example 6** the head office applies the primary rule under Recommendation 4 by aggregating the deductions claimed for actual expenditure and loss in each jurisdiction and comparing this against the total (tax adjusted) expenditures of the taxpayer. This adjustment has the effect of neutralising the mismatch associated with all duplicate expenditure claimed across the jurisdictions where the taxpayer operates. Similarly, in **Example 7**, the branch jurisdiction applies the secondary rule by comparing the aggregate tax deductions claimed for actual expenditure and loss in the branch and head office jurisdictions with the actual expenditures in those jurisdictions. This adjustment, under the secondary rule in Recommendation 4, has the effect of neutralising only those mismatch associated with all duplicate expenditure claimed in the relevant branch and head office jurisdiction.

Recommendation 4.3 – No branch mismatch to the extent set off against dual inclusion income

96. Recommendation 4.3 limits the operation of the double deduction rule to those cases where the payer jurisdiction permits the deduction to be set off against non-dual inclusion income.

97. Where the residence jurisdiction provides a general exemption from branch income then any deduction in the branch (that is also deductible in the residence jurisdiction) is likely to end up being set off against income that is not subject to tax in the residence jurisdiction. DD branch payments can also arise, however, in the context of taxable branches where the branch is permitted to join a tax group or there is some other mechanism in place in the branch jurisdiction that allows expenditure or loss to be set off against income derived by another person that is not taxable under the laws of the residence jurisdiction. A DD branch structure involving a taxable branch is illustrated in **Example 3** where the taxpayer restructures its branch operations and establishes a reverse hybrid entity to provide certain services to former branch customers. Another example of such a structure is illustrated in **Example 10** where the taxpayer establishes both a branch operation and an offshore subsidiary in a foreign jurisdiction that allows the subsidiary and the branch to form a group for tax purposes.

Timing of disallowance

98. Recommendation 6.3 of the Action 2 Report (OECD, 2015) requires an adjustment to be made under the deductible hybrid payments rule in those cases where the deduction may be set off against non-dual inclusion income in the payer jurisdiction. The Action 2 Report (OECD, 2015) states that is not necessary for a tax administration to know whether the deduction has actually been applied against non-dual inclusion income in the other jurisdiction before it is subject to restriction under the rule. The rules also, however, include a mechanism that allows jurisdictions to carry-forward deductions to a period where they can be set off against surplus dual inclusion income.

99. In certain cases the deductible hybrid payments rule may generate stranded losses by restricting a deduction in one jurisdiction even though the deduction that arises in the other jurisdiction cannot, in practice, be used to offset any income in that jurisdiction (because, for example, the business in that jurisdiction is in a net loss position). In this case Recommendation 6.1(d)(ii) of the Action 2 Report (OECD, 2015) provides that a tax administration may permit excess deductions to be set off against non-dual inclusion income where the taxpayer can establish that the deduction in the other jurisdiction cannot be offset against any income that is not dual inclusion income. The treatment of stranded losses is discussed in Example 6.2 of the Action 2 Report (OECD, 2015) where a taxpayer incurs losses in a foreign branch. In that example, the deductible hybrid payments rule has the potential to generate "stranded losses" if the taxpayer abandons its operations in the payer jurisdiction and winds up the branch at a time when it still has unused carry-forward losses from a prior period. The example notes that the tax administration may permit the taxpayer to set off any excess against non-dual inclusion income provided the taxpayer can establish that the winding up of the branch will prevent the taxpayer from using those losses anywhere else.

100. Denying (or restricting) the deduction at the time it arises (as contemplated under the Action 2 Report (OECD, 2015)) may have an unintended impact on direct investment through taxable branches and transparent entities. In particular, denying the taxpayer a

benefit of a loss suffered by the branch or hybrid entity until that taxpayer derives dual inclusion income may undermine one of the key tax objectives behind operating in branch form or through a tax transparent entity. Such a rule could discourage investment through foreign branches or transparent entities where losses may be incurred in early years. This issue could be addressed if the DD rule limited the deduction only to the extent the duplicate deduction was actually applied against non-dual inclusion income in the counterparty jurisdiction. This would mean that taxpayers with taxable branch operations (or investments through a transparent entity) could continue to deduct losses in respect of their offshore investment and that adjustments would only need to be made if and when the loss was used against non-dual inclusion income in the counterparty jurisdiction. It would also eliminate the need to allow for adjustments in respect of stranded losses.

101. Recommendation 4.3 accordingly provides that a double deduction will give rise to a branch mismatch only to the extent the payer jurisdiction allows the deduction to be set off against an amount that is not dual inclusion income. This ambiguity as to the timing of the disallowance gives the jurisdiction the flexibility to make the adjustment under the double deduction rule at the time the deduction arises (consistent with the treatment set out in Recommendation 6.3 of the Action 2 Report (OECD, 2015)) or at the time the deduction is actually offset against dual inclusion income under the laws of the payer jurisdiction. The domestic rules implementing the recommendations for neutralising DD outcomes in respect of hybrid entities and branches are likely to be the same (or similar) and jurisdictions may consider that any deferral of the adjustment under the DD rule that is permitted in respect of deductions claimed through a taxable branch, should also apply to DD outcomes arising through the use of a hybrid entity.

102. The difference in the timing of the adjustment under Recommendation 4 is illustrated in **Example 10**. In that example, a profitable parent company establishes both a subsidiary and a branch operation in another jurisdiction. The laws of that foreign jurisdiction permit the branch and the subsidiary to form a group for tax purposes. The branch incurs expenditure which results in net branch losses in the first two years of its operation. The branch then becomes profitable in the third year. Under the laws of the foreign jurisdiction these initial losses are partly available to be offset against the income of the subsidiary. The example illustrates the difference in the adjustments that could be made in order to give effect to the double deduction rule in the residence jurisdiction.

   a. Under the method set out in the Action 2 Report (OECD, 2015) (which requires an adjustment whenever the deduction may be set off against non-dual inclusion income in the foreign jurisdiction) the head office makes an adjustment under the laws of the residence jurisdiction for the full amount of the branch loss in each of the two years and carries the branch-loss forward to be set off against dual inclusion income of the branch in Year 3.

   b. Under the alternative method permitted under Recommendation 4.3 above (which requires an adjustment only when the payer jurisdiction allows the deduction to be set off against non-dual inclusion income) the head office can claim a portion of the branch loss in the initial period (to the extent it has not been used in the payer jurisdiction to offset income of the subsidiary) but is required to include additional amounts of income in subsequent years as the carry-forward loss in the payer jurisdiction is applied against non-dual inclusion income.

Note

1. Council Directive amending Directive (EU) 2016/1164 as regards hybrid mismatches with third countries dated 12 May 2017 ("ATAD 2").

Bibliography

Council of the European Union (2017), Council Directive amending Directive (EU) 2016/1164 as regards hybrid mismatches with third countries dated 12 May 2017 ("ATAD 2"), http://dsms.consilium.europa.eu/952/Actions/Newsletter.aspx?messageid=13108&customerid=37917&password=enc_643345636135526A32344361_enc, (accessed on 13 June 2017).

OECD (2015), *Neutralising the Effects of Hybrid Mismatch Arrangements, Action 2 – 2015 Final Report*, OECD Publishing, Paris, http://dx.doi.org/10.1787/9789264241138-en.

Chapter 5

Imported branch mismatch rule

1. Treatment of Imported Branch Mismatches

The payer jurisdiction should deny a deduction for any payment made under an imported branch mismatch arrangement to the extent that such payment directly or indirectly funds deductible expenditure under a branch mismatch arrangement.

2. Imported Branch Mismatch

An imported branch mismatch arrangement is a transaction or series of transactions that is entered into:

   a. between members of a controlled group; or

   b. as part of a structured arrangement to which the payer is a party,

that directly or indirectly funds deductible expenditure under a branch mismatch arrangement.

3. Limitation on Scope

This recommendation shall not apply to the extent that one of the jurisdictions involved in the transactions or series of transactions has made an equivalent adjustment in respect of such branch mismatch.

Overview

103. As described in the Introduction, a deductible payment can give rise to an imported branch mismatch where such payment directly or indirectly funds deductible expenditure under a branch mismatch arrangement. The policy behind the imported mismatch rule is to prevent taxpayers from entering into structured arrangements or arrangements with group members that shift the effect of an offshore branch mismatch into the domestic jurisdiction through the use of an instrument such as an ordinary loan.

104. Recommendation 5 of this report extends the scope of the imported mismatch rule in the Action 2 Report (OECD, 2015) to cover imported branch mismatches. This recommendation supplements, and does not replace, the imported mismatch recommendations in the Action 2 Report (OECD, 2015). It also uses language that is consistent with ATAD 2.[1]

105. Imported branch mismatches rely on the absence of effective branch mismatch rules in offshore jurisdictions in order to generate the mismatch in tax outcomes which can then be imported into the payer jurisdiction. The most reliable protection against imported branch mismatches will be for all jurisdictions to introduce branch mismatch rules recommended in this report. Such rules will neutralise the effect of the branch mismatch arrangement in the jurisdiction where the mismatch arises and prevent the effect of that mismatch being imported into a third jurisdiction.

106. The key objective of the imported branch mismatch rule is to maintain the integrity of the other branch mismatch rules by removing any incentive for multinational groups to enter into these arrangements. While these rules involve an unavoidable degree of co-ordination and complexity, they only apply to the extent a multinational group generates an intra-group deduction under a branch mismatch arrangement and will not apply to any payment that is made to a taxpayer in a jurisdiction that has implemented the full set of recommendations set out in this report.

107. In order to limit compliance costs and the risk of double taxation, each country that implements the recommendations set out in this report should make reasonable endeavours to implement an imported branch mismatch rule that adheres to the methodology set out in this report and to apply this methodology in the same way. This will allow the adjustments required under the imported mismatch rules in each jurisdiction to be calculated consistently for the whole group and in a way that avoids any unnecessary duplication of compliance obligations.

Recommendation 5.1 – Treatment of imported branch mismatches

Payment

108. The definition of payment used in Recommendation 5 is the same as that used for the other recommendations. A payment will only be treated as made under an imported branch mismatch arrangement if it is both deductible under the laws of the payer jurisdiction and gives rise to ordinary income under the laws of the payee jurisdiction. Payments will therefore include rents, royalties, interest and fees paid for services but will not generally include amounts that are treated as consideration for the disposal of an asset. A payment made to a person who is not a taxpayer in any jurisdiction will not be treated as an imported mismatch payment.

109. A payment should be treated as funding expenditure under a branch mismatch arrangement where the income from the payment is directly set off against a deduction under a branch mismatch arrangement or where the payment is indirectly set off against

that deduction through a chain of interconnected payments or group relief surrenders between intermediate taxpayers. A payment that is set off against a deduction under a deemed branch payment or a DD branch payment should not, however, be treated as having funded expenditure under an imported mismatch arrangement where that payment is treated as dual inclusion income.

110. This principle is illustrated in **Example 11** in the case of an intra-group payment made to a branch that is set off against a deemed branch payment. The example notes that the intra-group payment will not be subject to adjustment under Recommendation 5 if it was made to a taxable branch so that such payment is included in the income of both the residence and branch jurisdictions.

Tracing and priority rules

111. The guidance set out in the Action 2 Report (OECD, 2015) describes tracing and priority rules to be used by taxpayers and tax administrations to determine the extent to which a payment should be treated as set off against a deduction under an imported mismatch arrangement. These rules start by identifying the payment that gives rise to a hybrid mismatch (a "direct hybrid deduction") and then determine the extent to which that hybrid deduction has been funded (either directly or indirectly) out of payments made by taxpayers that are subject to the imported mismatch rule ("imported mismatch payments"). The same tracing and priority rules should be applied for determining the extent to which a payment directly or indirectly funds deductible expenditure under a branch mismatch arrangement (a "branch mismatch deduction").

112. In order to account for timing differences between jurisdictions and to prevent groups manipulating that timing in order to avoid the effect of the imported mismatch rule, a branch mismatch deduction should be taken to include any net loss that has been carried-forward to a subsequent accounting period, to the extent that loss results from a hybrid deduction. In order to reduce the complexity associated with the need to identify imported branch mismatches that arose prior to the publication of this report, any carry-forward loss from periods ending on or before 31 December 2016 should be excluded from the operation of this rule.

113. It will be the domestic taxpayer who has the burden of establishing, to the reasonable satisfaction of the tax administration, that the imported mismatch rule has been properly applied in that jurisdiction. This initial burden may be discharged by providing the tax administration with copies of the group calculations together with supporting evidence of the adjustments that have been made under the imported mismatch rules in other jurisdictions. Tax administrations will generally be relying on the taxpayer to provide them with these calculations and supporting evidence. In the absence of such information, a tax administration may consider issuing its own assessment of the extent to which income from an imported mismatch payment has been directly or indirectly set off against a branch mismatch deduction or hybrid deduction of another group member.

Recommendation 5.2 – Imported branch mismatch definition

114. The imported mismatch rule applies to both structured arrangements and imported mismatch arrangements that arise within a control group.

115. An imported branch mismatch arrangement should be treated as structured if the branch mismatch arrangement is structured and the deduction under the branch mismatch

and the imported mismatch payment form part of the same arrangement. The definition of arrangement is set out in Recommendation 12 of the Action 2 Report (OECD, 2015) and includes any agreement, plan or understanding and all the steps and transactions by which it is carried into effect. A structured imported mismatch arrangement therefore includes not only those payments and transactions that give rise to the branch mismatch but also all the other transactions and imported mismatch payments that are entered into as part of the same scheme, plan or agreement.

Recommendation 5.3 – Limitations on scope

116. As noted above, the most reliable protection against imported mismatches will be for jurisdictions to introduce hybrid and branch mismatch rules under the common approach set out in the Action 2. Such rules will address the effect of the hybrid or branch mismatch arrangement in the jurisdictions where it arises, and therefore prevent the effect of such mismatch being imported into a third jurisdiction. The imported mismatch rule therefore will not apply to any payment that is made to a taxpayer in a jurisdiction that has implemented the full set of Action 2 recommendations.

Note

1. See: Council Directive amending Directive (EU) 2016/1164 ("ATAD 2").

Bibliography

Council of the European Union (2017), Council Directive amending Directive (EU) 2016/1164 as regards hybrid mismatches with third countries dated 12 May 2017 ("ATAD 2"), http://dsms.consilium.europa.eu/952/Actions/Newsletter.aspx?messageid=13108&customerid=37917&password=enc_643345636135526A32344361_enc (accessed on 13 June 2017).

OECD (2015), *Neutralising the Effects of Hybrid Mismatch Arrangements, Action 2 – 2015 Final Report*, OECD Publishing, Paris, http://dx.doi.org/10.1787/9789264241138-en.

Annex A

Summary of recommendations

Recommendation 1 – Limitation to the scope of the branch exemption

> **1. Limitation to the scope of the branch exemption**
>
> Jurisdictions that provide an exemption for branch income should consider limiting the scope and operation of this exemption so that the effect of deemed payments, or payments that are disregarded, excluded or exempt from taxation under the laws of the branch jurisdiction, are properly taken into account under the laws of the residence jurisdiction.

Recommendation 2 – Branch payee mismatch rule

> **1. Denial of deduction for branch payee mismatches**
>
> The payer jurisdiction should deny a deduction for a payment that gives rise to a D/NI outcome to the extent that the mismatch is a result of:
>
> a. differences in the allocation of payments between the residence and the branch jurisdiction or between two branch jurisdictions; or
>
> b. the fact that the payment is to a disregarded branch.
>
> **2. Disregarded branch**
>
> A disregarded branch is a branch that is treated as giving rise to a taxable presence under the laws of the residence jurisdiction (and thus is eligible for an exemption from income) but is not treated as giving rise to a taxable presence under the laws of the branch jurisdiction.
>
> **3. Scope**
>
> This recommendation shall only apply to payments made under a structured arrangement or between members of a controlled group.

Recommendation 3 – Deemed branch payment rule

1. Denial of deduction for deemed branch payments

The jurisdiction that recognises a deemed branch payment (payer jurisdiction) should deny a deduction for that payment to the extent it gives rise to a branch mismatch.

2. Deemed branch payments

A deemed branch payment is a deemed payment between the branch and the head office or between two branches of the same taxpayer that gives rise to a D/NI outcome as a result of the fact that such payment is disregarded under the laws of the jurisdiction that is treated as receiving the payment (the payee jurisdiction).

3. No branch mismatch to the extent set off against dual inclusion income

A deemed branch payment shall give rise to a branch mismatch only to the extent the payer jurisdiction allows the deduction to be set off against an amount that is not dual inclusion income.

Recommendation 4 – Double Deduction Rule

1. Treatment of Double Deduction Outcomes

To the extent a double deduction outcome gives rise to a branch mismatch:

   a. the deduction should be denied in the investor jurisdiction; and

   b. where the deduction is not denied in the investor jurisdiction, then the deduction should be denied in the payer jurisdiction.

Any deduction should, however, be eligible to be offset against dual inclusion income whether arising in a current or subsequent period.

2. Double Deduction Outcome

A double deduction outcome means a deduction of the same payment, expense or loss in both the jurisdiction where such payment is made, expense is incurred or loss is suffered (the payer jurisdiction) and another jurisdiction (the investor jurisdiction).

3. No branch mismatch to the extent set off against dual inclusion income

A double deduction will give rise to a branch mismatch only to the extent the payer jurisdiction allows the deduction to be set off against an amount that is not dual inclusion income.

Recommendation 5 – Imported branch mismatch rule

1. Treatment of Imported Branch Mismatches

The payer jurisdiction should deny a deduction for any payment made under an imported branch mismatch arrangement to the extent that such payment directly or indirectly funds deductible expenditure under a branch mismatch arrangement.

2. Imported Branch Mismatch

An imported branch mismatch arrangement is a transaction or series of transactions that is entered into:

    a. between members of a controlled group; or

    b. as part of a structured arrangement to which the payer is a party,

that directly or indirectly funds deductible expenditure under a branch mismatch arrangement.

3. Limitation on Scope

This recommendation shall not apply to the extent that one of the jurisdictions involved in the transactions or series of transactions has made an equivalent adjustment in respect of such branch mismatch.

Annex B

Examples

Example 1	Branch payee mismatches
Example 2	Notional payment by taxable branch
Example 3	Taxable branch with non-dual inclusion income
Example 4	Notional payment by exempt branch
Example 5	Application of Recommendations 3 and 4 to notional payment
Example 6	Application of primary rule in Recommendation 4 to taxpayer with multiple branches
Example 7	Application of secondary rule in Recommendation 4 to taxpayer with multiple branches
Example 8	Allocation of third party expenses under Recommendation 3
Example 9	Allocation of third party expenses under Recommendation 4
Example 10	DD outcomes and treating mismatch as arising at the time of offset
Example 11	Imported mismatch

Example 1

Branch payee mismatches

Facts

1. In the example illustrated in the figure below, A Co (a company established and resident in Country A) establishes B Co, a software development company that is resident in Country B. B Co establishes a branch in Country C. B Co licences software to D Co (another group company) resident in Country D for use in its business of providing services to third party customers. D Co pays a deductible licence fee to B Co which is treated, under Country B law, as paid to the Country C Branch.

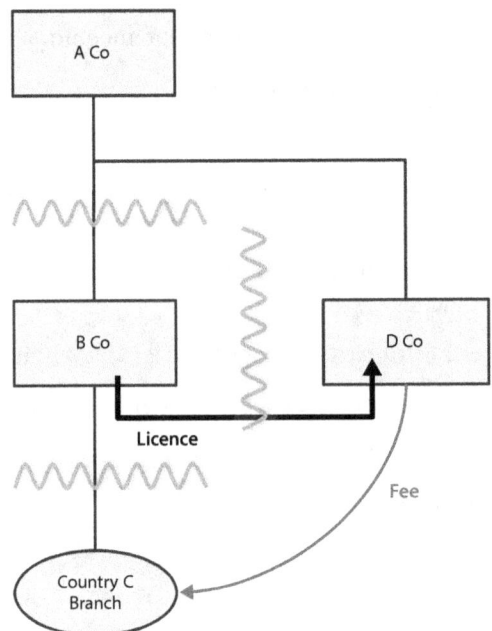

2. Country B provides an exemption for income derived by a foreign branch which means that the licence fee income is not subject to tax in Country B. The fee is not subject to tax in Country C. This creates an intra-group D/NI outcome. There could be a number of reasons why the payment of the licence fee is not subject to tax in Country C:

   a. Country C does not tax corporate taxpayers or only taxes residents of Country C.

   b. License fee income paid to the Country C Branch is eligible for a nil-rate of taxation under a special regime (such as a patent box).

   c. The branch does not give rise to a taxable presence for B Co under the domestic laws of Country C.

   d. The branch does not meet the legal definition of a permanent establishment under the Country B-C tax treaty, so that the payment received by the branch is excluded from Country C taxation under the relevant provisions of that treaty.

   e. Country C has different rules for allocating income to the branch or specific rules that exclude or exempt this type of income from taxation when paid to a non-resident.

Question

3. Does the mismatch identified in the arrangement above fall within any of the recommendations in this report?

Answer

4. The licence fee is not subject to tax in Country C. Accordingly, under Recommendation 1, Country B is encouraged (but not required) to consider narrowing the scope of its branch exemption to bring the licence fee into the charge to taxation. Having B Co take this payment into account for tax purposes under Country B law:

   a. will not necessarily trigger any additional Country B tax liability if the licence fee independently qualifies for an exemption from tax under the laws of Country B and

   b. may result in B Co recognising additional deductible expenditure under Country B law in connection with earning the licence fee.

5. In the case where the branch is treated as constituting a permanent establishment under the Country B-C tax treaty then the treaty may require Country B to exempt the licence fee from tax to the extent it is properly attributable to the branch and will prevent Country B from bringing the licence fee into ordinary income under Country B law.

6. In the event that there is no adjustment made in Country B under rules consistent with Recommendation 1 then Recommendation 2 shall apply to deny the deduction in Country D to the extent the payment gives rise to a D/NI outcome that is the result of a branch payee mismatch.

7. Therefore, the overall effect of the recommendations in this report is that:

   a. these type of payments should properly be subject to tax in the head office (if not included in income by the branch)

   b. if such payments are not included in income in any jurisdiction and the reason for this mismatch is either a result of:

      - a misallocation of that payment between the branch and the head office

      - the payment being made to a disregarded branch

      then a deduction for that payment should be denied where the payment is made intra-group or as part of a structured arrangement intended to produce a mismatch in tax outcomes.

Analysis

Country B should consider adjusting the scope of its branch exemption

8. Recommendation 1 of this report provides that jurisdictions, such as Country B, which exempt the income of foreign branches should consider narrowing the scope of this exemption so that it does not apply to payments that are not subject to tax under the laws of the branch jurisdiction. Recommendation 1 is not a branch mismatch rule, but rather a specific recommendation for changes to the scope and operation of the branch exemption in the residence jurisdiction intended to ensure that it does not have the effect of providing double taxation relief for payments that have not borne any tax.

9. Accordingly, under Recommendation 1, Country B is encouraged (but not required) to consider adjusting the scope and operation of the branch exemption to bring the licence fee into the charge to taxation under Country B law. Recommendation 1 could apply, not only in those cases where the reason for the mismatch is due to a misallocation of the payment under the laws of the branch jurisdiction, but also where the payment qualifies for tax-free treatment in the branch on some other basis.

10. There are a number of ways the residence jurisdiction could make an adjustment in order to include the payment in income under the laws of the residence jurisdiction that are consistent with a proper allocation of income and expenditure between the branch and the residence jurisdiction under agreed international standards. For example, Country B could expand the scope of its taxing regime to bring untaxed branch income into charge at the head office either by:

   a. requiring that any payment, which is derived by a resident taxpayer and not subject to tax in the branch jurisdiction, be brought into charge to taxation in the head office

   b. limiting the branch exemption to the amount of net income actually brought into the charge to tax by the branch.

11. In all cases, the adjustments required by Country B should be consistent with a proper allocation of income and expenditure between the branch and the residence jurisdiction and in line with the intended territorial scope of Country B's tax regime.

12. Requiring B Co to bring the licence fee into account in Country B under one of these methods will not automatically trigger an additional tax liability for B Co, if B Co can separately claim the benefit of a specific exemption for such payment under Country B law. Once brought into account under Country B law, for example, the licence fee could still be eligible for taxation at a nil or reduced rate, due to the fact that it relates to exploitation of intellectual property that is held subject to a preferential tax regime that is established under Country B law to encourage research and development (i.e. a "patent box" regime).

13. It is noted that in a case where the branch is treated as constituting a PE under the Country B-C tax treaty (and that treaty contains a provision equivalent to Article 23A of the Model Tax Convention) then the treaty may require Country B to exempt the licence fee from tax to the extent it is properly attributable to the branch, which will prevent Country B from bringing the licence fee into ordinary income under Country B laws.

Recommendation 2 only applies to the extent the payment gives rise to a D/NI outcome

14. A D/NI outcome arises where a payment is deductible under the laws of one jurisdiction and not included in ordinary income under the laws of any other jurisdiction. Although the licence fee may not be included directly in income by A Co it may be included in A Co's income under a CFC (or equivalent) regime. If Country D wishes to avoid the risk of economic double taxation from denying a deduction with respect to a licence fee that is, in fact, subject to tax under the CFC rules in Country A, then Country D should consider the extent of the adjustment required under the branch payee mismatch rule in light of such CFC inclusion. In this case D Co would need to satisfy the tax administration in Country D that the quantification and timing rules for the inclusion of CFC income under Country A law actually required that payment to be brought into account as ordinary income on A Co's tax return and D Co may be further required to demonstrate that the amount that is included does not carry an entitlement to any unrelated

Recommendation 2 only applies to the extent the D/NI outcome is the result of a branch payee mismatch

15. Whether Recommendation 2 applies to the facts of this example will also depend upon the reason why the payment is not subject to tax under Country C law. Recommendation 2 only applies to neutralise a D/NI outcome where the mismatch results from a payment to a disregarded branch or from differences in the allocation of payments between the residence and the branch jurisdictions. If the reason for the mismatch is because Country C does not impose corporate income tax or because the licence fee benefits from a preferential regime open to all taxpayers in Country C (such as a patent box regime) then the branch payee mismatch rules will not apply because the mismatch is not a result of any conflict in the allocation of payments between the branch and head office.

The fact that the payment is a diverted branch payment or a payment to a disregarded branch results in a D/NI outcome

16. If the Country C Branch does not give rise to a taxable presence under the domestic laws of Country C or does not meet the legal definition of a permanent establishment under the Country B-C tax treaty then Country C Branch may be considered a disregarded branch for tax purposes. Furthermore, if Country C law treats the licence fee as paid to the head office (or exempts or excludes the payment from tax on the grounds that the payment is made to a non-resident) then there is difference between Country B and Country C in the allocation of the licence fee and the licence fee should be treated as a diverted branch payment. In both cases the payment will be subject to adjustment under Recommendation 2 if it can be established that the mismatch is a result of the fact that the payment was a diverted branch payment or made to a disregarded branch.

17. As described in Chapter 2 of this report, this question can be answered by posing a counterfactual test that asks what the tax treatment of the payment would have been if it had been made directly to the head office. In this case the facts indicate that it is the operation of the branch exemption that shelters the licence fee from taxation under the laws of Country B, so that the payment would have been taxable if it was treated as paid to the head office. Accordingly, Recommendation 2 will operate to deny a deduction for the payment in the payer jurisdiction if the payment is to a disregarded branch or otherwise not subject to tax in the branch jurisdiction due to the fact that the same payment was treated as properly allocable to (and taxable in) the head office.

Recommendation 2 will not apply if Recommendation 1 applies

18. The disregarded branch or diverted branch payment rules will not apply, however, where the mismatch has been neutralised by a rule in Country B which ensures that a payment that is not brought into account in the branch must be brought into account in the head office. Thus if Country B, in accordance with Recommendation 1, restricts the scope of a branch exemption to payments that have actually been brought into the charge to taxation by the branch then the mismatch in tax outcomes would be neutralised and there should generally be no scope for the operation of the branch payee mismatch rule.

Example 2

Notional payment by taxable branch

Facts

1. A Co is a company that is established and tax resident in Country A. A Co provides computer services to customers located in Country A and B. Country B customers receive their services through a branch of A Co located in that country (i.e. Country B Branch).

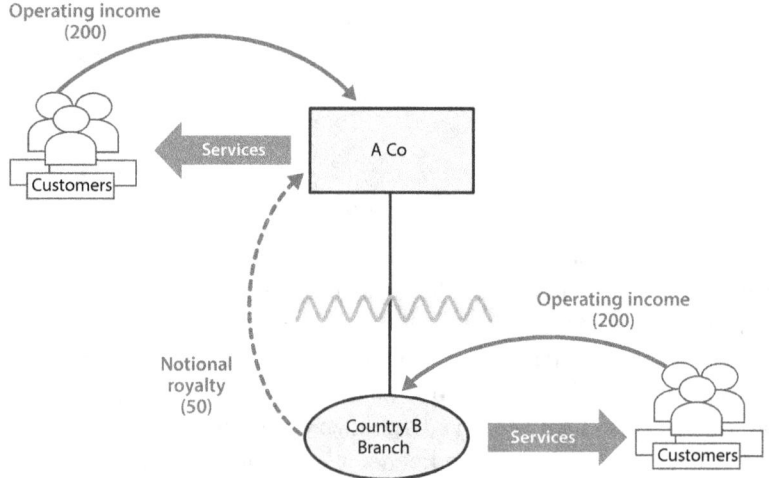

2. Under the laws of Country B, the income of the branch is fully taxable and the branch is permitted a deduction for a notional royalty payment made to the head office. This payment is intended to reflect an arm's length compensation for intellectual property that is owned by the head office and exploited by the branch in the course of providing services to Country B customers. The rules in Country A treat the income of the branch as fully taxable but do not recognise any notional payments between the branch and the head office.

Question

3. Does the notional royalty payment described above fall within Recommendation 3 of this report?

Answer

4. Recommendation 3 will not apply to adjust the deduction in respect of the notional royalty payment where the branch is treated as fully taxable under Country A law and the operating income of the branch exceeds the amount of the deemed payment.

Analysis

No branch mismatch if income of branch is fully taxable under Country A law

5. The deemed branch payment rule limits the ability of a taxpayer to set off a deduction from a deemed branch payment against non-dual inclusion income. However, in this case, where the amounts paid by Country B customers are treated as taxable income in both jurisdictions there is likely to be limited scope for the application of the deemed branch payment rule. Table B.2.1 provides an illustration of the position of the head office and the branch once all the branch income has been brought into account for tax purposes under Country A law.

Table B.2.1. **Taxable Branch**

Country A (Head Office)			Country B (Branch)		
Income	Tax	Book	Income	Tax	Book
Country A Customers	200	200			
Country B Customers	200	-	Country B Customers	200	200
Expenses			**Expenses**		
Employment	(40)	(10)	Employment	(30)	(30)
Administration costs	(40)	(20)	Administration costs	(20)	(20)
Research and development	(10)	(10)			
			Notional royalty payment	(50)	-
Net return before tax		160	Net return before tax		150
Taxable income	310		Taxable income	100	
Tax at 30%	(93)		Tax at 30%	(30)	
Credit	30				
Net tax to pay		(63)	Net tax to pay		(30)

6. As shown in Table B.2.1, A Co derives 200 of operating income from the respective computer service sales made in each of Country A and B and incurs 40 of administration costs (split evenly between the branch and the head office) and employment costs of 30 in the branch and 10 in the head office. The head office also recognises research and development expenses of 10 in respect of intellectual property (IP) that is used by the branch in providing services to customers. In total A Co has 400 of income and 90 of expenses leaving it with net income of 310 from its global operations.

7. A Co also has 310 of taxable income (because the full amount of the branch profits are taken into account under Country A law). Country A provides a full tax credit for the Country B tax imposed on income earned through the branch so that the final amount of tax payable under the laws of both jurisdictions is 30% of the net return.

8. Because all the operating income of the Country B Branch is taken into account as ordinary income under the laws of Country A, the deemed royalty payment recognised by the Country B Branch is deducted against dual inclusion income and no branch mismatch arises under Recommendation 3.

Example 3

Taxable branch with non-dual inclusion income

Facts

1. The facts are the same as in Example 2 except that in this case A Co restructures its operations in Country B by establishing a separate entity (B Co) to provide certain services to Country B customers that were previously supplied directly through the Country B Branch. B Co is a reverse hybrid (an entity that is treated as transparent under the laws of Country B but as a separate entity under Country A law). Country B branch continues to provide certain services to Country B customers after the restructure. A Co's and B Co's operations in Country B are illustrated in the figure below:

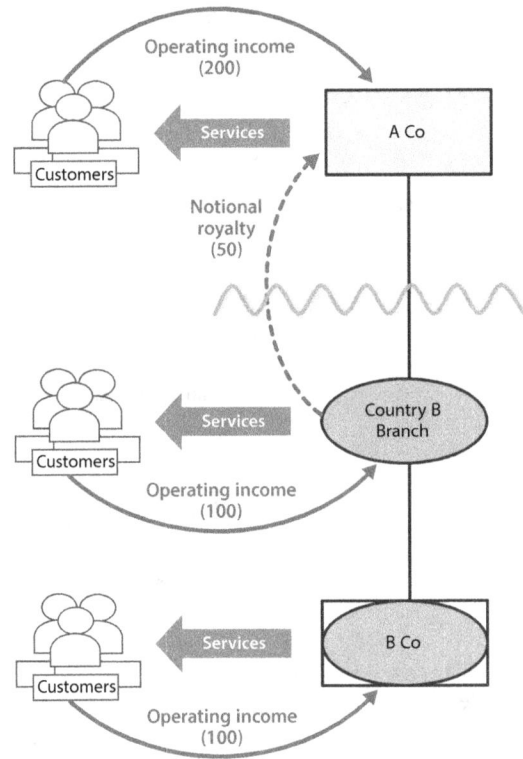

2. Following the restructuring, half of the operating income derived from Country B customers is now derived through B Co, which is a separate entity that is not subject to tax under Country A law. The total employment and administration expenses incurred in Country B are the same as in Example 2 but half of these expenses are now incurred by B Co. Country B Branch continues to claim a deduction for a deemed royalty payment paid to the head office and the amount of this deemed payment is the same as in Example 2.

3. Because B Co is disregarded under Country B law the income of B Co and the Country B Branch are treated as income of single entity so that the income and expenses of both the branch and company are recorded on a single tax return with any payments between them being disregarded for tax purposes. Table B.3.1 provides an illustration of the position of the head office and the Country B Branch following the restructuring:

Table B.3.1. **Taxable Branch with non-dual inclusion income**

Country A (A Co)			Country B (Country B Branch and B Co)		
Income	Tax	Book	Income	Tax	Book
Country A Customers	200	200			
Country B Customers	100	-	Country B Customers	200	200
Expenses			**Expenses**		
Employment	(25)	(10)	Employment	(30)	(30)
Administration costs	(30)	(20)	Administration costs	(20)	(20)
Research and development	(10)	(10)	Notional royalty payment	(50)	-
Net return before tax		160	Net return before tax		150
Taxable income	235		Taxable income	100	
Tax at 30%	(70.5)		Tax at 30%	(30)	
Credit	22.5*				
Net tax to pay		(48)	Net tax to pay		(30)

* Amount of credit may be subject to further limitation under Country A law.

4. There is no change to the overall tax position in Country B following the restructuring. All the tax payable under the laws of Country B is taxable at the level of the branch because B Co is not treated as a separate taxpayer for Country B tax purposes.

5. Under Country A law there is a decrease in the amount of Country B income and expenses included on A Co's return. A Co would ordinarily be expected to make a corresponding adjustment to the amount of foreign tax credits claimed in respect of its branch operations to reflect the fact that, following the restructuring, there are lower amounts of income and tax paid at the level of the branch.

Question

6. Will the notional royalty payment or any of the employment or administration expenses described above be subject to adjustment under the laws of the branch jurisdiction?

Answer

7. Following the restructuring, the dual inclusion income of the branch still exceeds the total amount of branch payments (including the deemed royalty payment and the employment and administration costs claimed in respect of the branch operations under both Country A and B law). Accordingly, the Country B Branch would not be expected to make any adjustment under the branch mismatch rules.

8. Country A could, however, consider applying rules that limit the amount of A Co's direct foreign tax credit to the (adjusted) net income of the branch after taking into account the effect of the notional royalty payment.

Analysis

No adjustments required under Country B law

9. The restructuring reduces the amount of income that is included under both Country A and B law, however the total amount of dual inclusion income still exceeds the amount of the deemed royalty payment and there is therefore no requirement for A Co to make an adjustment under Recommendation 3.

10. A Co might further consider whether any adjustment was required under Country B law in respect of the employment and administration costs claimed in both Country A and Country B (i.e. whether an adjustment is required in Country B under the double deduction rule in Recommendation 4.1(b). Again, however, no adjustment should be required under the branch mismatch rule because the branch is profitable on a stand-alone basis. The dual inclusion income of the branch exceeds the total amount of branch payments (including the deemed royalty payment of 50 and the 25 of employment and administration costs claimed in respect of the branch operations under both Country A and B law). Accordingly from Country B's perspective, branch payments do not exceed dual inclusion income and there should be no requirement to make any adjustment under the branch mismatch rules.

Calculation of direct foreign tax credits under Country A law

11. In this case Country A limits the amount of the foreign tax credit to the lesser of the amount of tax payable by the branch under Country B law and the marginal rate of tax under Country A law on branch income as calculated under Country A law. In this case the net income of the branch (as calculated under Country A law) is as follows:

Operating income	100
Employment costs	(15)
Admin costs	(10)
Net income (under Country A law)	75

The resulting limitation on the amount of direct foreign tax credits is therefore $(75 \times .30 =) 22.5$.

12. In this case, the effect of calculating the foreign tax credit using the principles governing the recognition of income and expenditure under Country A law, is that Country A does not take into account the impact of the D/NI outcome arising in respect of the deemed branch payment. This D/NI payment has the effect of reducing the amount of income subject to tax under Country B law without impacting on the calculation of the net income of the branch under Country A law.

13. While the branch mismatch rules do not directly impact on the amount of direct foreign tax credits a taxpayer may claim in respect of its branch operations, the calculation of these credits may give rise to tax policy concerns in the residence jurisdiction where they permit surplus tax relief to reduce or offset the tax on non-dual inclusion income. This issue could be addressed in Country A by limiting the amount of the direct foreign tax credit by reference to the (adjusted) net income of the branch, after taking into account the effect of the notional payments that have not been taken into account by the head office. In the absence of any such limitation in Country A, Country B may consider restricting the definition of dual inclusion income, so as not to include income that has been sheltered from tax by surplus foreign tax credits that have been recognised under Country A law.

Example 4

Notional payment by exempt branch

Facts

1. The facts are the same as in Example 2. A Co provides computer services to Country B customers through a branch located in that country (i.e. Country B Branch). In this case, however, Country A exempts branch income from taxation. A Co's operations in Country B are illustrated in the figure below:

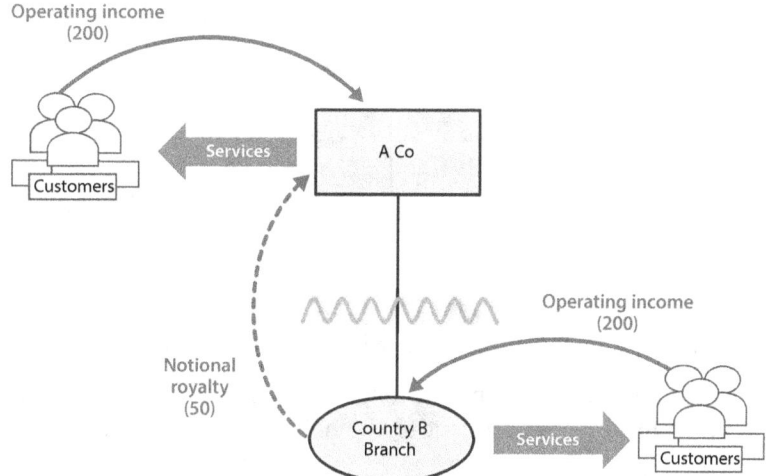

2. Under the laws of Country B, the income of the branch is fully taxable and the branch is permitted a deduction for a notional royalty payment made to the head office. This payment is intended to reflect an arm's length compensation for intellectual property that is owned by the head office and exploited by the branch in the course of providing services to Country B customers. The rules in Country A do not recognise any notional payments between the branch and the head office. Table B.4.1 provides an illustration of the position of the head office and the branch in respect of the deemed royalty payment.

3. The figures are the same as in Table B.2.1 except that the head office is only required to bring the income from its local operations in Country A into tax. The income derived by Country B Branch is exempt from tax under the laws of Country A. In this case, Country A denies a deduction for the research and development (R&D) expenses that would otherwise have been deductible under Country A law. This deduction is denied on the grounds that the intellectual property (IP) generated through such R&D is used solely in the Country B branch to derive income that is exempt from taxation under Country A law. As shown in the above Table B.4.1, A Co's net return (before tax) is 310 while the total taxable income under the laws of both jurisdictions is 270. The mismatch of 40 is the product of the D/NI outcome in respect of the notional royalty payment (50) adjusted by the denial of the R&D costs under Country A law (10).

Table B.4.1. **Exempt branch**

Country A			Country B		
Income	Tax	Book	Income	Tax	Book
Country A Customers	200	200	Country B Customers	200	200
Expenses			**Expenses**		
Employment	(10)	(10)	Employment	(30)	(30)
Administration costs	(20)	(20)	Administration costs	(20)	(20)
Research and development	-	(10)			
			Notional royalty payment	(50)	-
Net return before tax		160	Net return before tax		150
Taxable income	170		Taxable income	100	
Tax at 30%	(51)		Tax at 30%	(30)	
Net tax to pay		(51)	Net tax to pay		(30)

Questions

4. Does the notional royalty payment fall within Recommendation 1 of this report?

5. If there is no adjustment in Country A to take account of the notional royalty payment, is A Co required to make an adjustment to the net income of Country B Branch under Recommendation 3?

Answer

6. Recommendation 1 provides that Country A should consider making appropriate adjustments to the amount of income recognised by the head office so that the effect of any deemed payment made to the head office is properly taken into account under the laws of the residence jurisdiction. There are a variety of methods that Country A could adopt to eliminate the risk of mismatches arising in respect of notional payments. These methods may be less complicated than applying the deemed branch payment rule and may result in adjustments to items other than the deemed payment in order to properly reflect the allocation of income between the branch and the head office.

7. If Country A does not make an adjustment to properly reflect the notional royalty payment then A Co would be required to make an adjustment to the amount of net income recognised in the Country B Branch under Recommendation 3. This adjustment would take account of the fact that a portion of the deemed royalty payment had been recognised in Country A in the form of a denial of the deduction for research and development expenses in respect of IP assets that have been allocated to the branch.

Analysis

Application of Recommendation 1

8. Recommendation 1 provides that Country A should consider making modifications to the scope and operation of its branch exemption so that the effect of any deemed payments made to the head office are properly taken into account under the laws of the residence jurisdiction. This report does not set out any limitations on the amount of the adjustment or provide any detail on the most appropriate mechanism for making that adjustment provided it remains consistent with the relevant tax treaty obligations and tax policy settings in that jurisdiction. One example of the type of adjustment that could be made in the residence jurisdiction is shown in Table B.4.1 above where Country A has denied a deduction for certain R&D expenses associated with an IP asset that has been used in the branch to generate exempt income. This denial of an "equivalent category of expenditure" as described below, could be considered as one way in which the residence jurisdiction takes into account the effect of the deemed payment by the branch. The residence jurisdiction could adopt other methods for recognising additional income in the head office jurisdiction in an amount equal to the deemed payment.

Recognising the deemed payment as an item of additional income

9. Country A could, for example, introduce a rule requiring taxpayers in the position of A Co to include the deemed payment made by an exempt branch as ordinary income. This type of adjustment is illustrated in Table B.4.2.

Table B.4.2. **Exempt branch recognising deemed payment in payee jurisdiction**

Country A			Country B		
Income	Tax	Book	Income	Tax	Book
Country A Customers	200	200	Country B Customers	200	200
Expenses			**Expenses**		
Employment	(10)	(10)	Employment	(30)	(30)
Administration costs	(20)	(20)	Administration costs	(20)	(20)
Research and development	(10)	(10)			
			Notional royalty payment	(50)	-
Adjustment	50				
Net return before tax		160	Net return before tax		150
Taxable income	210		Taxable income	100	
Tax at 30%	(63)		Tax at 30%	(30)	
Net tax to pay		(63)	Net tax to pay		(30)

10. The mismatch in tax outcomes is eliminated by the head office recognising the amount of the deemed branch payment in income. Country A has also permitted the head office to make a corresponding adjustment to the deductibility of the R&D expenses in order to properly reflect the fact that the underlying IP asset is now treated as giving rise to taxable income in the residence jurisdiction (in the form of the adjustment for the deemed payment).

Granting head office a deduction for the net income of the branch

11. A deemed branch payment will not give rise to a mismatch where the rules for calculating branch income in the residence jurisdiction operate in such a way as to ensure that the scope of the branch exemption only covers income that is subject to tax in the branch jurisdiction. Table B.4.3 illustrates an alternative mechanism for calculating branch income which limits the scope of the exemption to the amount of income that is actually subject to tax in the branch jurisdiction. This methodology ensures that any income that is sheltered by the deemed royalty payment will be subject to tax in the head office.

Table B.4.3. **Exempt branch with deduction for branch income**

Country A			Country B		
Income	Tax	Book	**Income**	Tax	Book
Country A Customers	200	200			
Country B Customers	200	200	Country B Customers	200	200
Expenses			**Expenses**		
Employment	(40)	(10)	Employment	(30)	(30)
Administration costs	(40)	(20)	Administration costs	(20)	(20)
Research and development costs	(10)	(10)			
Deduction for net branch income	(100)		Notional royalty payment	(50)	-
Net return before tax		160	Net return before tax		150
Taxable income	210		Taxable income	100	
Tax at 30%	(63)		Tax at 30%	(30)	
Net tax to pay		(63)	Net tax to pay		(30)

Application of Recommendation 3

12. If A Co does not make an adjustment that takes into account the payment of the deemed royalty under rules consistent with Recommendation 1, then A Co should consider the extent to which Recommendation 3 applies to neutralise the mismatch in tax outcomes under the laws of Country B.

13. The deemed branch payment rule limits the ability of a taxpayer to set off the deduction from a deemed branch payment against non-dual inclusion income when such payment is not included in income by the payee.

Notional royalty is a deemed payment

14. In this example, the notional royalty payment falls within the definition of a deemed payment under Recommendation 3 as it is a notional payment between the branch and head office that does not represent (and is not calculated by reference to) an actual expenditure of the taxpayer. While, in this case, A Co's accounts do recognise expenditure on R&D, the facts do not indicate that the notional royalty payment (or any part of the payment) has been calculated by reference to those R&D costs. The R&D expenditure is not the same type of outgoing as a notional royalty payment. The former is in respect of the development of an

IP asset while the latter is a payment for use of that IP asset. It would therefore be difficult to trace, with precision, the notional royalty payment into the R&D expense such that it can be reliably determined that both items are (in reality) deductions for the same expense. Accordingly, it cannot be said that the R&D expenditure recognised in A Co's accounts is itemised expenditure which can directly be attributed to the notional royalty payment.

Deemed payment is disregarded (other than to the extent it is recognised by an allocation of expenditure or loss of an equivalent category)

15. A deemed payment will not give rise a mismatch unless it is "disregarded" under the laws of the payee jurisdiction. The head office may recognise a deemed payment by including it directly in income or by allocating expenditure or loss of an equivalent category to the payer jurisdiction.

16. In this case (and as illustrated in Table B.4.1) Country A limits the deductibility of the R&D expense on the grounds that the resulting IP asset is used in deriving exempt branch income. Where the payee jurisdiction has domestic rules limiting the deductibility of expenditure on the basis that such expenditure is allocable to the branch then the effect of this limitation should be taken into account in determining the extent to which a deemed payment has been disregarded under the deemed branch payment rule.

17. In this case, the deemed payment and allocated expenditure pertain to the same general category of assets (being the IP used in providing services to customers) and the basis on which the R&D expense has been denied in the residence jurisdiction indicates that there is a straightforward connection between the deemed payment and the allocated expenditure or loss. Accordingly, these two items should be treated as being in an equivalent category for the purposes of the deemed branch payment rule. It is noted that the deemed payment (a royalty for the use of an IP asset) does not need to be of the same specific type as the expenditure or loss allocated by the payee (R&D costs) and does not need to be calculated on the same basis. However, a deemed payment should only be treated as recognised by the allocation of an equivalent category of expenditure or loss to the extent of the amount actually allocated to the payer jurisdiction and that the expenditure or loss has been denied in the residence jurisdiction as a result of such allocation.

Mismatch is a result of the payment being disregarded

18. A branch mismatch only arises where the D/NI outcome is a result of the fact that the deemed payment is disregarded under the laws of the payee jurisdiction. This is a counterfactual test that asks what the tax treatment of the payment would have been if it had been recognised by A Co. In this case the facts indicate that A Co is a taxable entity so the resulting mismatch is one that arises as a result of the deemed payment. Table B.4.4 shows the position of the head office and the branch following the adjustment under Recommendation 3.

19. In total A Co has 400 of income and 90 of expenses leaving it with net income of 310 from its global operations. All the operating income of the Country B Branch is exempt from tax under the laws of Country A so that the deemed royalty payment recognised by the Country B Branch is deducted against non-dual inclusion income. The deemed royalty payment is not properly taken into account under the laws of Country A so that Country B denies a deduction for the amount of the deemed royalty except to the extent that Country A has allocated an equivalent category of expenditure to the branch in the form of a denial of a deduction for the R&D expenses.

Table B.4.4. **Adjustment under Recommendation 3 for Exempt Branch**

Country A			Country B		
Income	Tax	Book	Income	Tax	Book
Country A Customers	200	200	Country B Customers	200	200
Expenses			**Expenses**		
Employment	(10)	(10)	Employment	(30)	(30)
Administration costs	(20)	(20)	Administration costs	(20)	(20)
Research and development	-	(10)			
			Notional royalty payment	(50)	-
			Rec. 3 Adjustment	50	
			Research and development costs (allocated by head office)	(10)	
Net return before tax		160	Net return before tax		150
Taxable income	170		Taxable income	140	
Tax at 30%	(51)		Tax at 30%	(42)	
Net tax to pay		(51)	Net tax to pay		(42)

20. The overall impact of the recommendations in this report is that, if the head office does not properly take into account the effect of the notional payment, the jurisdiction that allows for the notional payment should not provide a deduction for such payment to the extent that the income or expenses associated with that payment are not taken into account under the laws of the payee jurisdiction.

21. The net effect of these rules is to ensure that the deduction for the deemed payment is only available when (and to the extent) the taxpayer has taken the effect of that payment into account in the counter-party jurisdiction. An adjustment under the branch mismatch rules ensures that the full amount of the taxpayer's net income is brought into charge under the laws of either the branch or the residence jurisdiction while ensuring that the adjustments do not result in double taxation.

Example 5

Application of Recommendations 3 and 4 to notional payment

Facts

1. The facts of this example are the same as in Example 4 except that, in addition, A Co contracts for various services from third party service providers. A Co's operations (including the service fees paid to third party service providers) are illustrated in the figure below.

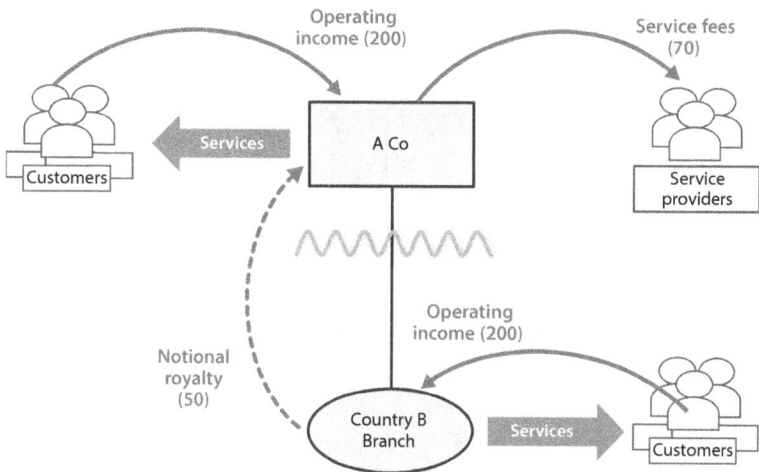

2. Table B.5.1 provides an illustration of the position of the head office and the branch. The figures in the table below are the same as those in Table B.4.1 except that:

   a. the head office recognises an additional 70 of third party expenses for accounting purposes (only 50 of which is deductible under Country A law)

   b. the Country B Branch treats 30 of these third party expenses as incurred directly by the branch.

3. As in Table B.4.1, A Co derives 200 of operating income from the respective computer service sales made in each of Country A and B and incurs 40 of administration costs (split evenly between the branch and the head office) and employment costs of 30 in the branch and 10 in the head office. Country A also denies a deduction for the research and development (R&D) costs on the grounds that the intellectual property (IP) generated by such R&D is used solely to derive exempt income in Country B Branch.

4. In total A Co has 400 of income and 160 of expenses leaving it with net income of 240 from its global operations, however the net effect of the allocation of the R&D costs, deemed royalty and additional third party expenditure between the branch and head office is that A Co only recognises 190 of taxable income across both jurisdictions (i.e. a mismatch of 50).

Table B.5.1. **Mismatch arising in respect of deemed and actual payments**

Country A			Country B		
Income	Tax	Book	Income	Tax	Book
Country A Customers	200	200	Country B Customers	200	200
Expenses			**Expenses**		
Employment	(10)	(10)	Employment	(30)	(30)
Administration costs	(20)	(20)	Administration costs	(20)	(20)
Research and development	-	(10)			
Third party services	(50)	(70)			
			Notional royalty payment	(50)	-
			Third party services	(30)	-
Net return before tax		90	Net return before tax		150
Taxable income	120		Taxable income	70	

Question

5. How do the recommendations in this report apply to neutralise the mismatch in tax outcomes arising from the use of this structure?

Answer

6. A Co should apply Recommendation 3 before determining the amount of any double deduction subject to adjustment under Recommendation 4.

7. If the notional payment can accurately and reliably be traced through to an item of expenditure of the same type recorded in the taxpayer's accounts then the Country B Branch should treat the notional payment (to that extent) as an actual payment of the underlying expenditure incurred by A Co. The balance of the notional royalty payment that does not directly relate to actual expenditure of the taxpayer should be subject to adjustment under Recommendation 3.

8. Consistent with the analysis set out in Example 4, Country B should deny a deduction for the amount of the deemed royalty except to the extent that Country A has allocated an equivalent category of expenditure to the branch in the form of a denial of a deduction for R&D costs.

9. There is still a mismatch in tax outcomes under the branch structure following the application of the deemed branch payment rule in Country B. This mismatch arises due to the fact the branch and the head office are claiming deductions for third party services that exceed, in aggregate, the actual amount of expenditure on these services (i.e. a double deduction outcome). This mismatch will be subject to adjustment in Country A under the primary rule in Recommendation 4.

Analysis

Apply Recommendation 3 (deemed branch payments rule) before Recommendation 4 (DD rule)

10. The mismatch that arises in this example is due to differences in the allocation of actual and deemed expenditure between various parts of the same enterprise. There are two recommendations dealing with mismatches that arise in these circumstances:

    a. Recommendation 3 which requires the branch jurisdiction to deny a deduction for a deemed payment to the extent such payment is disregarded by the head office.

    b. Recommendation 4 which requires the residence jurisdiction to deny a deduction to the extent the same expense is deductible under the laws of the branch jurisdiction.

11. Consistent with the Action 2 Report (OECD, 2015), the taxpayer should apply Recommendation 3 before determining the amount of any double deduction subject to adjustment under Recommendation 4. This, in turn, requires A Co to determine the extent to which any deduction claimed by Country B Branch represents an allocation of an actual expense of the taxpayer.

Apply Recommendation 3 to notional royalty payment to the extent such payment does not represent an allocation of third party expenses

12. In the previous example, the notional royalty payment was treated as a deemed payment under Recommendation 3 because it did not represent (and was not calculated by reference to) an actual expenditure of the taxpayer. In this example, however, the facts indicate that A Co has incurred additional expenses in respect of third party services and it is possible that some of these expenses can be directly attributed to the notional royalty payment recognised by Country B Branch.

13. A notional payment should be treated as an actual payment where the payment relates to specific functions performed, assets held or risks assumed by another part of the same taxpayer and there is itemised expenditure in the tax accounts of the payee of the same type that can be directly attributed to the notional payment. Assume, for example, that part of the third party services supplied to A Co includes information technology (IT) licences and support services that relate to software owned by A Co (which, in turn, forms part of the basis for the notional royalty paid by Country B Branch). Assume further that these third party services are charged on a per-user basis so that A Co can determine (without the need to collect any further information or perform complex calculations) the portion of the expenditure that is attributable to Country B Branch.

14. In this case, even though the notional royalty payment is not expressly calculated by reference to such third party services, the notional payment can be defined with sufficient precision such that it can be traced through to an item of expenditure of the same type recorded in the payee's accounts and the nature of that expenditure is such that it can reliably and directly be attributed to the deemed payment. If this is the case then Country B Branch should treat the notional payment (to that extent) as an actual payment of the underlying expenditure incurred by A Co.

15. Table B.5.2 indicates the position of A Co under Country A and Country B law following the adjustment required under Recommendation 3.

16. In this case A Co can determine that the notional royalty payment treated as made by Country B Branch is attributable (in part) to software owned by A Co and that a portion

of the third party services expenditure is directly attributable to the costs of using (and supporting the use) of that software. Furthermore there is itemised expenditure in A Co's management accounts that accurately allows a portion of these third party service costs (10) to be attributed to the activities of the Country B Branch. Accordingly A Co treats a portion of the notional royalty payment as actual third party expenditure on software and support services.

Table B.5.2. **Adjustment under Recommendation 3**

Country A			Country B		
Income	Tax	Book	Income	Tax	Book
Country A Customers	200	200	Country B Customers	200	200
Expenses			**Expenses**		
Employment	(10)	(10)	Employment	(30)	(30)
Administration costs	(20)	(20)	Administration costs	(20)	(20)
Research and development	-	(10)	Deemed Royalty	(40)	-
Third party services	(50)	(70)	Software license and IT support	(10)	-
			Other third party services	(30)	-
			Rec. 3 Adjustment	40	
			Research and development costs (allocated by head office)	(10)	-
Net return before tax		90	Net return before tax		150
Taxable income	120		Taxable income	100	

17. The balance of the notional royalty payment does not directly relate to actual expenditure of the taxpayer and should be subject to adjustment under Recommendation 3. Consistent with the analysis set out in Example 4, Country B should deny a deduction for the amount of the deemed royalty except to the extent that Country A has allocated an equivalent category of expenditure to the branch in the form of a denial of a deduction for the R&D costs. The allocation of deductible expenditure should only be treated as equivalent to an adjustment under Recommendation 3 where the head office has been denied a deduction for such expenditure due to the fact that such amount has actually been allocated to the payer jurisdiction. In this case, the facts indicate that R&D costs are entirely attributable to the intellectual property used by Country B Branch and therefore should be wholly taken into account by the Country B Branch when determining the amount of the adjustment under Recommendation 3.

Remaining mismatch is attributable to DD outcome and subject to adjustment under the primary rule in Recommendation 4

18. Following the application of the deemed branch payment rule there is still a mismatch in tax outcomes under the branch structure. This is because both the branch and the head office are claiming deductions for third party services and those deductions exceed, in aggregate, the actual amount of expenditure on these services.

19. It may be possible for A Co to identify, on an item by item or category by category basis, the extent to which the amount of deductible expenditure claimed in the branch jurisdiction exceeds the amount that has been allocated to the branch by the head office. In more complex commercial branch operations, however, it will generally be impractical for a taxpayer to undertake this kind of detailed analysis. In these cases A Co should be permitted, under the laws of the relevant jurisdiction, to adopt a simpler implementation solution for tracking double deductions and dual inclusion income that is based, as much as possible, on existing domestic rules, administrative guidance, presumptions and tax calculations while still meeting the basic policy objectives of Recommendation 4.

20. For example, under Country A law, A Co could determine the total amount of double deductions on an aggregate basis by comparing the deductions claimed for actual expenditure and loss in the branch and head office jurisdictions against the taxpayer's total expenditures (excluding those expenditures that were not deductible under the laws of either the branch or head office jurisdiction). This excess may be treated as a double deduction (subject to adjustment under Recommendation 4) to the extent it cannot be explained solely by reference to differences in timing or valuation. An example of this calculation, based on the figures in Table B.5.2 is set out below.

Table B.5.3. **Calculation of total deductions claimed in Branch and Head Office**

Deductions for actual expenditure under Country A law		
Employment	(10)	
Administrative costs	(20)	
Third party services	(50)	
Total actual deductible expenditures (Country A)		(80)
Deductions for actual expenditure under Country B law		
Employment	(30)	
Administrative costs	(20)	
Software license and IT support	(10)	
Other third party services	(30)	
Research and development costs*	(10)	
Total actual deductible expenditures (Country B)		(100)
Total tax deductions under both jurisdictions in relevant period		(180)

* Note that when taking into account aggregate deductions for expenditure in the Country B Branch, the branch should take into account any reduction in the adjustment made under Recommendation 3 due to an allocation of equivalent expenditure by the head office.

21. The total tax deductions claimed on the branch and head office return for the relevant period exceed the actual (tax adjusted) expenditure in the accounts. The difference of 20 (which is not attributable to differences in the timing in the recognition of expenditure) should be treated as giving rise to a double deduction. Table B.5.4 sets out the adjustment required under both Country A and Country B law.

Table B.5.4. **Adjustments under Recommendations 3 and 4**

Country A			Country B		
Income	Tax	Book	Income	Tax	Book
Country A Customers	200	200	Country B Customers	200	200
Expenses			**Expenses**		
Employment	(10)	(10)	Employment	(30)	(30)
Administration costs	(20)	(20)	Administration costs	(20)	(20)
Research and development	-	(10)			
Third party services	(50)	(70)			
			Deemed Royalty	(40)	-
			Software license and IT support	(10)	-
			Other third party services	(30)	-
Rec. 4 Adjustment	20		Rec. 3 Adjustment	40	
			Research and development costs (allocated by head office)	(10)	
Net return before tax		90	Net return before tax		150
Taxable income	140		Taxable income	100	

22. Therefore the overall impact of the recommendations in this report on the facts of this example is that:

   a. The branch jurisdiction should not allow a deduction for a notional payment to the extent that the income or expenses associated with that payment are not taken into account under the laws of the residence jurisdiction.

   b. Any deductions for actual expenditure that are taken into account in both the head office and the branch are denied at the level of the head office to the extent the branch has already set those deductions off against (exempt) branch income.

23. The net effect of these rules is to ensure that the branch only grants a deduction for a deemed payment when (and to the extent) that the taxpayer has taken the effect of that payment into account in the counter-party jurisdiction and that the total of the tax deductions claimed by the taxpayer in the branch and head office do not exceed the taxpayer's actual deductible expenditure. Adjustments under the branch mismatch rules ensure that the full amount of the taxpayer's net income is brought into charge under the laws of either the branch or the residence jurisdiction while ensuring that the adjustments do not result in double taxation.

ANNEX B. EXAMPLES – **79**

Example 6

Application of primary rule in Recommendation 4 to taxpayer with multiple branches

Facts

1. The facts of this example are the same as in Example 5 except that A Co also has an identical branch in Country C. Country A exempts the income of both branches from taxation. As in the previous example, A Co contracts for various services from third party service providers. Both Country B and Country C Branches recognise a notional payment (or payments) to the head office to compensate the head office for the performance of services or the assumption of risks or ownership of assets held by the head office. A Co's operations in Country B and Country C (including the service fees paid to third party service providers) are illustrated in the figure below.

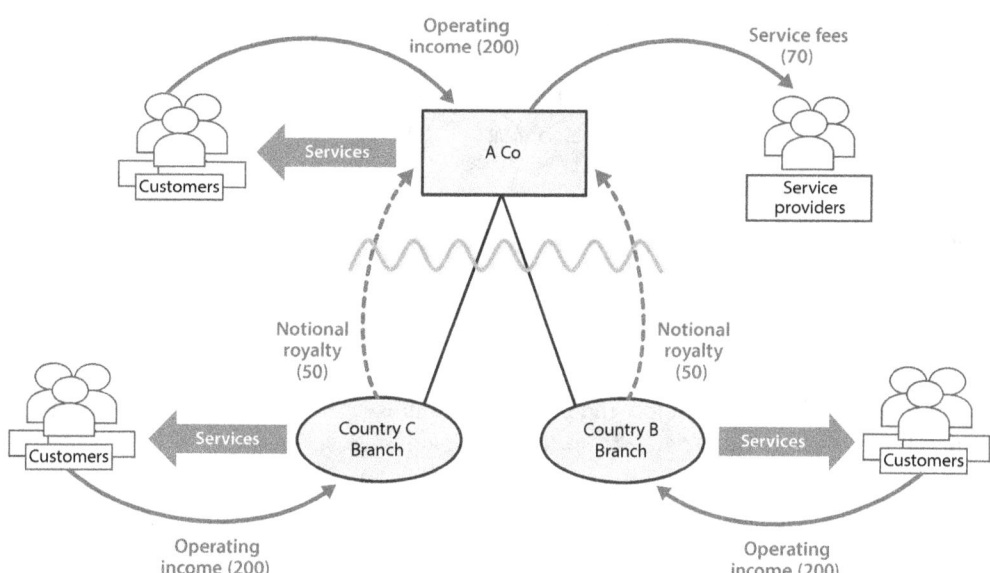

2. In this case it is assumed that Country B has applied the deemed branch payment rule to neutralise the mismatch arising in respect of the notional payment between Country B Branch and the head office. Table B.6.1 provides an illustration of the net position of the head office and the branches for tax purposes.

3. The figures for Countries A and B set out in Table B.6.1 are the same as those shown in B.5.2 except that A Co is only permitted to deduct 30 out of its total expenditure on third party services of 70 (the balance of the expenditure being treated as allocated evenly between the two branches). While A Co is ordinarily entitled to deduct research and development (R&D) costs, this deduction is denied owing to the fact that the intellectual property (IP) in question is used in Country B branch (see the analysis in Example 4 above).

4. The branch operations in Country B are the same as those described in Example 5 (and the adjustment made under the deemed branch payment rule is therefore the same as

set out in that example). While the branch operations in Country C are the same as those in Country B, Country C has not implemented the branch mismatch rules and therefore does not make any adjustment under Recommendation 3 in respect to the deemed branch payment. The net effect of these allocations (after the application of the deemed payment rule in Country B) is that A Co is required to include an aggregate of 310 of taxable income against a net return (before tax) of 390 (i.e. there is a mismatch of 80).

Table B.6.1. **Mismatch arising in respect of deemed and actual payments**

Country A			Country B			Country C		
Income	Tax	Book	Income	Tax	Book	Income	Tax	Book
Country A Customers	200	200	Country B Customers	200	200	Country C Customers	200	200
Expenses			**Expenses**			**Expenses**		
Employment	(10)	(10)	Employment	(30)	(30)	Employment	(30)	(30)
Administration costs	(20)	(20)	Administration costs	(20)	(20)	Administration costs	(20)	(20)
Research and development	-	(10)	Deemed Royalty	(40)	-	Deemed Royalty	(40)	-
Third party services	(30)	(70)	Software license and IT support	(10)	-	Software license and IT support	(10)	-
			Other third party services	(30)	-	Other third party services	(30)	-
			Rec. 3 Adjustment	40				
			Research and development costs (allocated by head office)	(10)				
Net return before tax		90	Net return before tax		150	Net return before tax		150
Taxable income	140		Taxable income	100		Taxable income	70	

Question

5. How would Country A apply the primary rule in Recommendation 4 to neutralise the mismatch in tax outcomes arising from the arrangement described above?

Answer

6. To the extent the mismatch is attributable to double deduction outcomes it will be subject to adjustment in Country A under the primary rule in Recommendation 4. Recommendation 4 will not, however, operate to neutralise the mismatch associated with the deemed royalty payment made by the Country C Branch.

Analysis

Adjustment under primary rule in Recommendation 4

7. Under the primary rule in Recommendation 4 the investor jurisdiction (Country A) should restrict the deductibility of any payment, expense or loss that is also deductible under the laws of the payer jurisdictions (Countries B and C) so that such amount can only be set off against income that is dual inclusion income. In this case, where Country A provides a general exemption in respect of branch income then any deduction in the branch jurisdiction that is also deductible in the residence jurisdiction is likely to end up being set off against income that is not subject to tax in the residence jurisdiction.

8. As in Example 5, it will generally be impractical to expect A Co to undertake a line-by-line (or even category-by-category) investigation into whether the amount of deductible expenditure claimed in the branch jurisdiction exceeds the amount that has been allocated to the branch by the head office and A Co should be permitted, under the laws of Country A, to use an implementation solution that is simple, robust and based, as much as possible, on existing domestic rules, administrative guidance, presumptions and tax calculations while still meeting the basic policy objectives of Recommendation 4.

9. For example, A Co could determine the total amount of double deductions on an aggregate basis by comparing the deductions claimed for actual expenditure and loss in the branch and head office jurisdictions against the taxpayer's total expenditures (excluding those expenditures that were not deductible under the laws of either the branch or head office jurisdiction). This excess may be treated as a double deduction (subject to adjustment under Recommendation 4) to the extent it cannot be explained solely by reference to differences in timing or valuation. An example of this calculation, based on the figures in Table B.6.2, is set out below.

Table B.6.2. **Calculation of total deductions claimed in each jurisdiction**

Deductions for actual expenditure under Country A law		
Employment	(10)	
Administrative costs	(20)	
Third party services	(30)	
Total actual deductible expenditures (Country A)		(60)
Deductions for actual expenditure under Country B law		
Employment	(30)	
Administrative costs	(20)	
Software license and IT support	(10)	
Other third party services	(30)	
Research and development costs (allocated to B Branch)*	(10)	
Total actual deductible expenditures (Country B)		(100)
Deductions for actual expenditure under Country C law		
Employment	(30)	
Administrative costs	(20)	
Software license and IT support	(10)	
Other third party services	(30)	
Total actual deductible expenditures (Country C)		(90)

* Note that when taking into account aggregate deductions for expenditure in the Country B Branch, the branch should take into account any reduction in the adjustment made under Recommendation 3 due to an allocation of equivalent expenditure by the head office.

10. Table B.6.2 above shows that the total deductions claimed in the branch and head office tax returns for the relevant period is 250. This total should be compared with 210 of total expenditure recorded in the accounts for tax purposes resulting in an excess of 40. This excess may be treated as a double deduction (subject to adjustment under Recommendation 4) to the extent it cannot be explained solely by reference to differences in timing or valuation.

Table B.6.3. **Adjustments under Recommendations 3 and 4**

Country A			Country B			Country C		
Income	Tax	Book		Tax	Book	Income	Tax	Book
Country A Customers	200	200	Country B Customers	200	200	Country C Customers	200	200
Expenses			**Expenses**			**Expenses**		
Employment	(10)	(10)	Employment	(30)	(30)	Employment	(30)	(30)
Administration costs	(20)	(20)	Administration costs	(20)	(20)	Administration costs	(20)	(20)
Research and development	-	(10)	Deemed Royalty	(40)	-	Deemed Royalty	(40)	-
Third party services	(30)	(70)	Software license and IT support	(10)	-	Software license and IT support	(10)	-
			Other third party services	(30)	-	Other third party services	(30)	-
Rec. 4 Adjustment	40		Rec. 3 Adjustment	40				
			Research and development costs (allocated by head office)	(10)				
Net return before tax		90	Net return before tax		150	Net return before tax		150
Taxable income	180		Taxable income	100		Taxable income	70	

11. Table B.6.3 sets out the adjustment required under Country A law. Note that this adjustment does not entirely eliminate the mismatch in tax outcomes and the remaining mismatch is attributable to the recognition of the notional payment in Country C.

12. The overall impact of the recommendations in this report on this arrangement is that:

   a. Country B does not allow a deduction for a notional payment to the extent that the income or expenses associated with that payment are not taken into account under the laws of the residence jurisdiction.

   b. Any deductions for actual expenditure that are taken into account in both the head office and the branch are denied at the level of the head office to the extent the branch has already set those deductions off against (exempt) branch income.

   c. No adjustment is required in respect of the branch mismatch that is attributable to the notional payment between the Country C Branch and the head office because Country C has not adopted the branch mismatch rules.

13. The net effect of these rules is to ensure that the branch jurisdiction (Country B) only grants a deduction for the deemed payment when (and to the extent) that the taxpayer has taken the effect of that payment into account in the counter-party jurisdiction. In addition, the application of Recommendation 4 in Country A ensures that the total of the tax deductions claimed by the taxpayer in the branch and head office do not exceed the taxpayer's actual deductible expenditure. In respect of the outstanding mismatch, this example illustrates that changes to the scope of the branch exemption under Recommendation 1 may, in certain cases, be a more effective mechanism for addressing branch mismatches than making adjustments at the level of the branch.

Example 7

Application of secondary rule in Recommendation 4 to taxpayer with multiple branches

Facts

1. The facts of this example are the same as in Example 6 except that only Countries B and C have implemented the branch mismatch rules.

Question

2. How does the deemed branch payment rule in Recommendation 3 and the secondary rule in Recommendation 4 apply to neutralise the mismatch in tax outcomes arising from this arrangement?

Answer

3. Countries B and C should deny a deduction for the amount of the deemed royalty payment, under the deemed branch payment rule in Recommendation 3 except to the extent that Country A has allocated an equivalent category of expenditure to the Country B Branch in the form of a denial of a deduction for research and development (R&D) costs. Each branch should also be denied a deduction under the double deduction rule in Recommendation 4 to the extent the aggregate tax deductions claimed in the branch and head office jurisdiction exceed the total amount of (tax adjusted) expenditure in the head office and each branch.

Analysis

Adjustments required under Country B law

4. Consistent with the analysis set out in Example 4, Country B should deny a deduction for the amount of the deemed royalty except to the extent that Country A has allocated an equivalent category of expenditure to the branch in the form of a denial of a deduction for R&D costs.

5. As in Example 6, it may be impractical for the Country B Branch to undertake a line by line comparison of each item of income and expenditure to determine whether a double deduction has arisen in the branch. Country B may allow the branch to aggregate the tax deductions claimed for actual expenditure and loss in the branch and head office and compare this amount against actual expenditures in order to determine the amount of double deductions that are claimed by the branch.

6. An example of the calculation of the expenditures for the head office and Country B Branch is set out in Table B.6.2. This total (160) can be compared with total expenditure recorded in the accounts for the branch and head office (i.e. the actual expenditures of the branch and head office adjusted to reflect any amounts that are treated by the head office as allocable to another branch). Table B.7.1 sets out the amount of actual expenditure incurred by the Country B Branch and head office.

Table B.7.1. **Calculation of expenditure in each jurisdiction**

Actual expenditure of head office		
Employment	(10)	
Administrative costs	(20)	
Research and development (allocated to Country B Branch)	(10)	
Third party services	(70)	
Adjustment for amount allocated to Country C Branch	20	
Total actual expenditures (Country A)		(90)
Actual expenditure of Country B Branch		
Employment	(30)	
Administrative costs	(20)	
Total actual expenditures (Country B)		(50)
Total expenditure in head office and Country B Branch		(140)

7. There are a number of methods that a taxpayer could use to calculate the amount of expenditure incurred in the head office and branch jurisdiction. The method that is used by the taxpayer should be based on the existing accounts as prepared in accordance with the relevant standards in the jurisdictions where the taxpayer operates. In the calculation set out in Table B.7.1 above, A Co has started with the expenditures recorded in the head office accounts, and then adjusted these amounts for expenditure that can properly be treated as attributable to the Country C branch, in order to determine the total expenditure in the head office and Country B Branch. In making this calculation the taxpayer will be expected to adopt a simple but reliable and consistent methodology for making such adjustments that would be capable of being used in each jurisdiction that applies rules consistent with Recommendation 4. In this case where there is already an allocation of third party expenses under the laws of Country A (and it is assumed that Country B and C Branches are identical), A Co has divided that allocation evenly between the two jurisdictions.

8. The excess of deductions claimed over actual expenditure in this case is (160 - 140 =) 20. This excess may be treated as a double deduction (subject to adjustment under Recommendation 4) to the extent it cannot be explained solely by reference to differences in timing or valuation.

Adjustments required under Country C law

9. A similar calculation can be made under Country C law. The required adjustment under Recommendation 3 is greater than in Country B, however, as the adjustment is not offset by an allocation of equivalent expenditure in the form of R&D costs. Under Recommendation 4, the total deductions claimed for actual expenditure in both jurisdictions is 150 and the total expenditure recorded in the accounts for the branch and head office (adjusted to reflect any amounts that are treated by the head office as allocable to the Country B Branch) is 130. The excess of deductions claimed over actual expenditure in this case is (150 - 130 =) 20. This excess of 20 may be treated as a double deduction (and subject to adjustment under Recommendation 4) to the extent it cannot be explained solely by reference to differences in timing or valuation. Table B.7.2 sets out the net amount of adjustment required under Country B and C law.

Table B.7.2. **Adjustments under Country B and C law**

Country A			Country B			Country C		
Income	Tax	Book		Tax	Book	Income	Tax	Book
Country A Customers	200	200	Country B Customers	200	200	Country C Customers	200	200
Expenses			**Expenses**			**Expenses**		
Employment	(10)	(10)	Employment	(30)	(30)	Employment	(30)	(30)
Administration costs	(20)	(20)	Administration costs	(20)	(20)	Administration costs	(20)	(20)
Research and development	-	(10)	Deemed Royalty	(40)	-	Deemed Royalty	(40)	-
Third party services	(30)	(70)	Software license and IT support	(10)	-	Software license and IT support	(10)	-
			Other third party services	(30)	-	Other third party services	(30)	-
			Rec. 3 adjustment	40		Rec 3 Adjustment	40	
			Research and development costs (allocated by head office)	(10)				
			Rec. 4 adjustment	20		Rec. 4 adjustment	20	
Net return before tax		90	Net return before tax		150	Net return before tax		150
Taxable income	140		Taxable income	120		Taxable income	130	

10. Table B.7.2 sets out the adjustments required under Country B and C law. Note that these adjustments eliminate the mismatch in tax outcomes owing to the fact that the deemed branch payment has been neutralised in Country C under Recommendation 3. The overall impact of the recommendations in this report on this arrangement is that:

   a. Country B Branch and Country C Branch do not allow a deduction for a notional payment to the extent that the income or expenses associated with those payments are not taken into account under the laws of the residence jurisdiction; and

   b. Any deductions for actual expenditure that are taken into account in both the head office and the branch are denied at the level of the branch to the extent the head office has set those deductions off against income that is not taxable in the branch.

11. The net effect of these rules is to ensure that the branch only grants a deduction for a deemed payment when (and to the extent) that the taxpayer has taken the effect of that payment into account in the counter-party jurisdiction and that the total of the tax deductions claimed by the taxpayer in the branch and head office do not exceed the taxpayer's actual deductible expenditure. Adjustments under the branch mismatch rules ensure that the full amount of the taxpayer's net income is brought into charge under the laws of either the branch or the residence jurisdiction while ensuring that the adjustments do not result in double taxation.

Example 8

Allocation of third party expenses under Recommendation 3

Facts

1. A Co is a company established and resident in Country A. A Co uses its own equity and money borrowed from an unrelated bank to make a loan to a customer located in Country A (Customer A). A Co also lends funds to a customer located in Country B (Customer B) through a branch established in that country (Country B Branch).

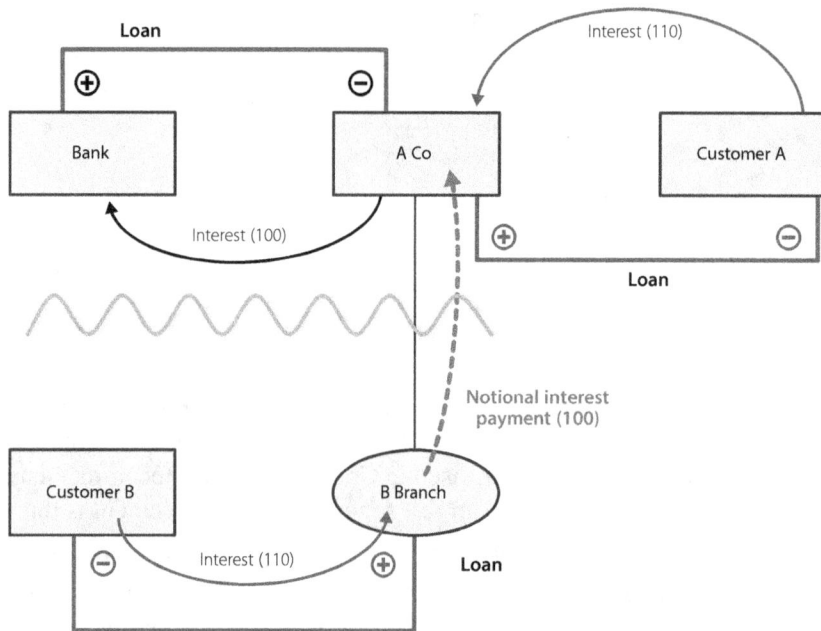

2. The rules in Country A for allocating income and expenditure to the branch require the head office to treat a portion of the interest paid to the bank as attributable to the exempt branch (and that portion is therefore non-deductible under Country A law).

3. Country B law calculates the net income of Country B Branch as if it was a separate entity for tax purposes, however, in making this calculation, Country B treats the branch as making an interest payment to the head office. Table B.8.1 illustrates the mismatch in tax outcomes that arises under this structure.

4. As shown in Table B.8.1, A Co earns a total of 220 of interest income and has 100 of interest expenses. The net return (before tax) under the arrangement is therefore 120. Under Country B law, the branch is treated as taxable on the interest payment of 110 from Customer B and is entitled to a deduction for the notional interest expense of 100 on a hypothetical loan from the head office. The net income subject to tax in Country B is therefore 10.

5. Under Country A law, the head office of A Co is also treated as deriving 110 of taxable interest income. The interest paid by Customer B is eligible for the branch exemption and not subject to tax under Country A law. A Co is, however, required to allocate half the interest expense on the bank loan to the exempt branch for tax purposes so that the total amount of interest that is deductible under Country A law is only 50 leaving the head office with net taxable income under Country A law of 60.

6. The overall effect of this arrangement is that while A Co's net return under the arrangement is 120, A Co only has total taxable income of 70 under the laws of Country A and B.

Table B.8.1. **Mismatch arising from notional payment**

Country A			Country B		
	Tax	Book		Tax	Book
Income			**Income**		
Interest from Customer A	110	110			
			Interest from Customer B	110	110
Expenditure			**Expenditure**		
Interest paid to bank	(50)	(100)	Notional interest deduction	(100)	
Net return		10	Net return		110
Taxable income	60		Taxable income	10	

Question

7. How do the recommendations in this report apply to neutralise the mismatch in tax outcomes arising from this structure?

Answer

8. The notional interest payment treated as made by B Branch should be treated as an actual interest expense to the extent the payment represents or is calculated by reference to actual interest expenditure recognised in the accounts of the payee. The effect of treating the notional payment as an actual interest expense is that the mismatch will be subject to adjustment under Recommendation 4.

9. If the notional interest payment cannot be accurately and reliably traced through to an actual item of interest expenditure recognised in the taxpayer's accounts then Country B should deny a deduction for the amount of the deemed interest payment except to the extent that Country A has allocated an equivalent category of financing costs to the branch.

Analysis

Notional payment subject to adjustment under Recommendation 4 to the extent it represents an actual interest expense

10. While this payment is treated, under the laws of Country B, as a notional interest payment to the head office, if, in practice, the payment is calculated by reference to A Co's actual borrowing costs (or the interest expenditure or borrowing costs in the tax accounts of the payee that can be directly attributed to the notional interest payment) then the interest expense claimed under Country B law should not be treated as a deemed payment for the purposes of the deemed branch payments rule. This type of notional interest payment is (in reality) an allocation of third party interest costs to the branch under Country B law which should be treated as giving rise to a branch DD outcome subject to adjustment under Recommendation 4 (see the supporting analysis in Example 9).

11. The facts of this example involve only one loan and a single notional interest expense. In branch financing operations of any significant size a taxpayer may have some difficulty in tracing notional interest expenses to actual third party borrowing costs. In the context of significant financing operations, the taxpayer may have entered into a number of borrowing, security and hedging transactions that will make it difficult (if not impossible) to trace the notional interest charge to any identifiable third party expense. In these cases, where the notional payment is not expressly calculated by reference to actual expenditure of the payee, and there is no itemised expenditure of the same type in the tax accounts of the payee that can be directly attributed to that notional payment, then the taxpayer should treat the notional payment as a deemed payment subject to adjustment under Recommendation 3.

No adjustment required under Recommendation 3 to the extent head office allocates expenditure of an equivalent category

12. A deemed interest payment between the branch and the head office is not subject to adjustment under the deemed branch payment rule to the extent the payment is recognised through an actual allocation of third party interest expense by the head office under Country A law.

13. Unlike the tracing approach described above, which is used to determine whether a notional payment represents or is calculated by reference to actual expenditure of the taxpayer, the determination of whether a deemed payment belongs to an equivalent category as an item of expenditure or loss in the head office jurisdiction is a broader test that should be done on a like-kind basis. In this case, both the deemed payment recognised under Country B law and the expenditure required to be allocated under Country A law relate to the same general category of expenditure (i.e. financing costs) and accordingly the two items should be treated as being in an equivalent category for the purposes of the deemed branch payment rule.

14. The deemed payment does not need to be of the same type as the expenditure or loss allocated by the payee and does not need to be calculated on the same basis. Accordingly, if the financing costs in the payee jurisdiction that were allocated to the branch included swap, derivative or guarantee fees they should still be treated as expenditure of an equivalent category despite the fact that they are of a different type and calculated on a different basis.

15. In this case, therefore, a portion of the notional interest treated as paid by the branch to the head office under Country B law (50) is treated as recognised in the residence jurisdiction in the same period by virtue of the corresponding allocation made by the head office to the branch under the laws of Country A. No adjustment would be required under the deemed branch payment rule to the extent the notional payment (under Country B law) is treated as recognised by this allocation. The deemed branch payment rule will continue to apply, however, to the extent the notional interest paid to head office was not recognised through a corresponding allocation of third party interest. Accordingly, in this example, only a portion (50) of the notional interest expense would be caught by the deemed branch payment rule.

16. The overall impact of the recommendations in this report on this arrangement is that:

   a. A notional payment that is, in reality, an allocation of third party interest costs to the branch should be treated as giving rise a double deduction outcome potentially subject to adjustment under Recommendation 4.

   b. A notional payment that cannot be attributed to any third party expense (i.e. a deemed payment) is not deductible in the branch if that payment exceeds the amount of deductible expenditure of an equivalent category allocated to the payer jurisdiction by the branch.

17. The net effect of these rules is to ensure that the branch only grants a deduction for a deemed payment when (and to the extent) that the taxpayer has taken the effect of that payment into account in the counter-party jurisdiction. Adjustments under the branch mismatch rules ensure that the full amount of the taxpayer's net income is brought into charge under the laws of either the branch or the residence jurisdiction while ensuring that the adjustments do not result in double taxation.

Table B.8.2. **Adjustment under Recommendation 3**

Country A			Country B		
	Tax	Book		Tax	Book
Income			**Income**		
Interest from Customer A	110	110			
			Interest from Customer B	110	110
Expenditure			**Expenditure**		
Interest paid to bank	(50)	(100)	Notional interest deduction	(100)	
			Recommendation 3 Adjustment	50	
Net return		10	Net return		110
Taxable income	60		Taxable income	60	

Example 9

Allocation of third party expenses under Recommendation 4

Facts

1. The facts are the same as those in Example 8 except that there is no notional interest payment between the branch and the head office. A Co uses its own equity and money borrowed from an unrelated bank to make a loan to a customer located in Country A (Customer A) and a customer located in Country B (Customer B). The loan to Customer B is made through a branch established in that country (Country B Branch). The structure of A Co's lending operations are illustrated in the figure below.

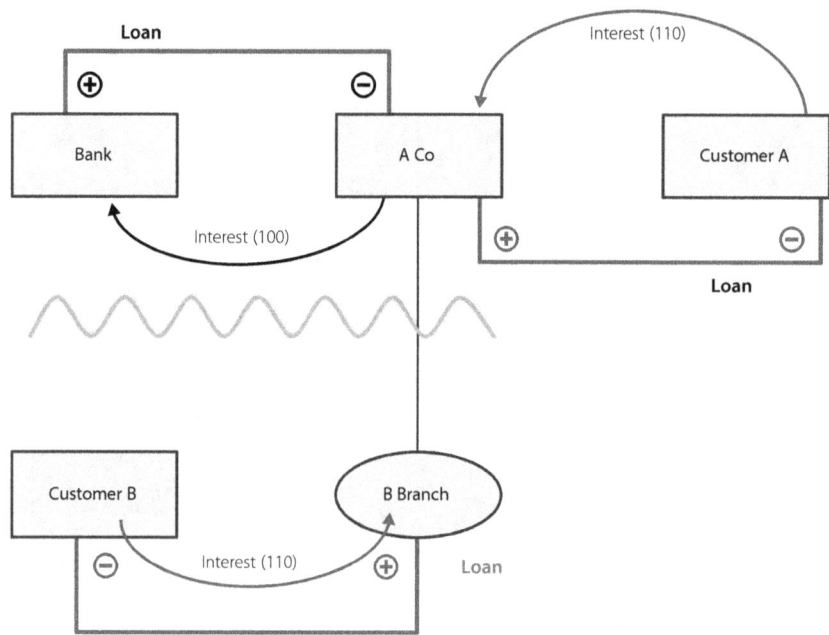

2. The rules in Country A for allocating income and expenditure to the branch require the head office to treat a portion of the interest paid to the bank as attributable to the exempt branch (and that portion is therefore non-deductible under Country A law). Country B law calculates the net income of Country B Branch as if it was a separate entity for tax purposes, however, in making this calculation, Country B applies a tracing approach to interest deductibility which treats all of the interest expenditure incurred by A Co as attributable to the branch. Table B.9.1 illustrates the mismatch in tax outcomes that arise under this structure.

3. As shown in Table B.9.1, A Co earns a total of 220 of interest income and has 100 of interest expenses. The net return (before tax) under the arrangement is therefore 120. Under Country B law, the branch is treated as taxable on the interest payment of 110 from Customer B and is entitled to a deduction for all the interest expense incurred on the loan from the bank (100). The net income subject to tax in Country B is therefore 10.

Table B.9.1. **Mismatch arising from double deduction**

	Country A			Country B		
	Tax	Book		Tax	Book
Income			**Income**		
Interest from Customer A	110	110			
			Interest from Customer B	110	110
Expenditure			**Expenditure**		
Interest paid to bank	(50)	(100)	Interest paid to bank	(100)	
Net return		10	Net return		110
Taxable income	60		Taxable income	10	

4. Under Country A law, the head office of A Co is also treated as deriving 110 of taxable interest income. The interest paid by Customer B is eligible for the branch exemption and not subject to tax under Country A law. A Co is, however, required to allocate half the interest expense on the bank loan to the exempt branch for tax purposes so that the total amount of interest that is deductible under Country A law is only 50, leaving the head office with net taxable income under Country A law of 60.

5. As in Example 8, the overall effect of this arrangement is that while A Co's net return under the arrangement is 120, A Co only has total taxable income of 70 under the laws of Country A and B.

Question

6. How does Recommendation 4 of the branch mismatch report apply to neutralise the mismatch in tax outcomes arising from this structure?

Answer

7. Under the primary rule in Recommendation 4, Country A should restrict the deductibility of any interest expense that is also deductible under the laws of Country B. A similar adjustment should be made in Country B under the secondary rule where the deduction is not denied in the residence jurisdiction.

Analysis

8. A double deduction outcome arises where the same payment, expense or losses deductible in the jurisdiction where such payment is made, expenses are incurred or losses are suffered (the payer jurisdiction) and another jurisdiction (the investor jurisdiction). In this case where the actual interest expenditure is treated as incurred directly through the branch, it is the branch jurisdiction that should be treated as the payer jurisdiction and the residence jurisdiction as the investor jurisdiction.

9. Recommendation 4 applies to neutralise a double deduction outcome to the extent it gives rise to a branch mismatch. Recommendation 4.1 requires the adjustment to first be made in the investor jurisdiction (in this case, at the level of the head office). Recommendation 4.3 provides that no mismatch will arise to the extent that a deduction is set off against an amount that is included in income under the laws of both the investor

and the payer jurisdictions (i.e. dual inclusion income). In this case, however, because of the operation of the branch exemption in Country A, none of B Branch's income is subject to tax in Country A in the relevant period.

Application of the primary response

10. In this case it is the residence jurisdiction that should apply the primary response. Country A should deny A Co's duplicate deductions to the extent it gives rise to a mismatch in tax outcomes. Table B.9.2 sets out the required adjustment under the rule.

11. The head office would be entitled to carry the denied interest deduction forward in accordance with its ordinary domestic rules and this deduction would be available to be set off against future dual inclusion income. Such dual inclusion income could arise, for example, where the rules for allocating income and expense to the branch and head office resulted in the same item of income being treated as taxable under the laws of both jurisdictions.

Table B.9.2. **Adjustment under Recommendation 4.1 (a)**

	Country A			Country B	
	Tax	Book		Tax	Book
Income			**Income**		
Interest from Customer A	110	110			
Interest from Customer B	-	110	Interest from Customer B	110	-
Expenditure			**Expenditure**		
Interest paid to bank	(50)	(100)	Interest paid to bank	(100)	-
Adjustment	50				
Net return		120	Net return		-
Taxable income	110		Taxable income	10	

Application of the defensive rule

12. In the event Country A does not apply the primary response, Country B should deny a deduction for the payment to the extent necessary to prevent that deduction from being set off against income that is not dual inclusion income. The total amount of adjustment required under Country B law would be calculated as set out in Table B.9.3.

Table B.9.3. **Adjustment under Recommendation 4.1 (b)**

	Country A			Country B	
	Tax	Book		Tax	Book
Income			**Income**		
Interest from Customer A	110	110			
Interest from Customer B	-	110	Interest from Customer B	110	-
Expenditure			**Expenditure**		
Interest paid to bank	(50)	(100)	Interest paid to bank	(100)	-
			Adjustment	50	
Net return		120	Net return		-
Taxable income	60		Taxable income	60	

13. The overall impact of the recommendations in this report on this arrangement is that any deductions for actual expenditure that are taken into account in both the head office and the branch are denied at the level of the head office (or the branch) to the extent the counterparty jurisdiction allows the deduction to be set off against non-dual inclusion income. The structure and ordering of the rules in Recommendations 3 and 4 ensures that the mismatch is neutralised without giving rise to the risk of double taxation.

Example 10

DD outcomes and treating mismatch as arising at the time of offset

Facts

1. The facts of this example are the same as that illustrated in Figure 4 of this report. A Co has established both a branch operation and a subsidiary in Country B. Country B law permits the subsidiary (B Co) and the Country B Branch to form a group for tax purposes, which allows the expenditure incurred by the Country B Branch to be offset against the income of the subsidiary.

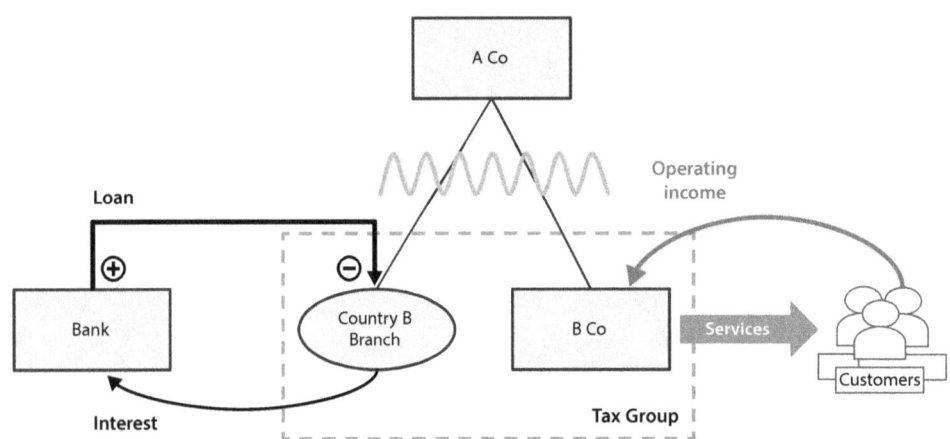

2. The net income (and loss) positions of A Co, Country B Branch and B Co over a 3 year period are as follows:

Table B.10.1. **Net income (and loss) positions**

	Year 1	Year 2	Year 3	Total
A Co (excluding branch)	800	800	800	2 400
B Branch	(400)	(200)	300	(300)
B Co	200	300	400	900
Total	600	900	1 500	3 000

3. If Country B Branch is treated as taxable under the laws of Country A, then the interest expense incurred by the branch will give rise to separate deductions under the laws of Country A and Country B. A Co will claim the deduction directly on the head office tax return, while this interest expenditure can also be offset, under Country B law, against the operating income derived by the subsidiary (i.e. against non-dual inclusion income). This structure therefore permits the same interest expense to be set off simultaneously against different items of income in the residence and branch jurisdiction.

4. The expected tax outcome in Country A (assuming that both countries apply tax at a marginal rate of 30%) will be as set out in Table B.10.2.

Table B.10.2. **Expected tax outcomes in Country A**

	Year 1	Year 2	Year 3	Total
Net income under Country A law	400	600	1 100	2 400
Tax under Country A law	(120)	(180)	(330)	(630)
Direct foreign tax credit			60	60
Total tax paid				(570)

5. As set out in Table B.10.2, the net income of A Co includes the expense incurred by the branch in the first two years. In Year 3 the branch turns a profit (perhaps due to the fact that a portion of the outstanding loan is forgiven by the bank) resulting in net income in the branch of 300.

6. It is assumed, for these purposes, that A Co will be entitled to a direct foreign tax credit for tax paid on branch income (as calculated under the laws of Country A). Accordingly the net tax paid in Country A over the three year period (taking into account the foreign tax credit) will be 570.

7. The expected tax outcome in Country B will be as follows:

Table B.10.3. **Expected tax outcomes in Country B**

	Year 1	Year 2	Year 3	Total
Net income under Country B law	(200)	100	700	600
Apply loss carry-forward		(100)	(100)	
Income subject to tax	0	0	600	
Tax under Country B law			(180)	(180)
Total tax paid				(180)

8. As set out in Table B.10.3, the combined net income of Country B Branch and B Co includes a deduction for the interest expense incurred by the branch in the first two years. This results in no net income and carry-forward losses for the first two years in respect of Country B's operations. In Year 3, the branch and company both become profitable resulting in net income under Country B law of 700 and income subject to tax (after the application of carry-forward losses) of 600.

9. The net effect of this structure is, therefore, that the group has both a net return and taxable income from its global operations of 3000. However, the effect of taking into account the foreign tax credit under Country A law is that A Co only pays tax of 750 on this income (out of an expected tax burden of 900).

Question

10. How does Recommendation 4 of the branch mismatch report apply to neutralise the mismatch in tax outcomes arising from this structure?

Answer

11. Recommendation 4.3 provides that a double deduction will give rise to a branch mismatch only to the extent the payer jurisdiction allows the deduction to be set off against an amount that is not dual inclusion income. Jurisdictions should have the flexibility to make the adjustment under the double deduction rule either at the time the deduction arises or at the time the deduction is actually offset against dual inclusion income under the laws of the payer jurisdiction.

Analysis

12. Recommendation 4 provides that a double deduction will give rise to a branch mismatch only to the extent the payer jurisdiction allows the deduction to be set off against an amount that is not dual inclusion income. This ambiguity as to the timing of the disallowance gives the jurisdiction the flexibility to make the adjustment under the double deduction rule either:

   a. at the time the deduction arises (following the treatment set out in Recommendation 6.3 of the Action 2 Report (OECD, 2015))

   b. at the time the deduction is actually offset against dual inclusion income under the laws of the payer jurisdiction.

Adjustments provided for under Action 2 Report (OECD, 2015)

13. Table B.10.4 sets out the required adjustment under the primary rule in Recommendation 4 adopting the timing rules set out in Recommendation 6.3 of the Action 2 Report (OECD, 2015).

Table B.10.4. **Adjustment under Recommendation 4**

	Year 1	Year 2	Year 3	Total
Net income under Country A law	400	600	1 100	2 400
Adjustment under Rec. 4	400	200	(300)	
	800	800	800	
Tax under Country A law	(240)	(240)	(240)	(720)
Direct foreign tax credit	0	0	0	0
Total tax paid				(720)

14. The net income that would otherwise be included under Country A law is adjusted by the application of the primary rule in Recommendation 4. In the first two years there is a reduction in the amount of the deduction claimed through the branch (owing to the fact that in both these years the deduction may be set off against non-dual inclusion income). In year 3 the branch derives 300 of dual inclusion income and the branch loss that has been carried-forward is offset against the dual inclusion income in that year.

15. Although the branch income is subject to tax in Country B in year 3, Country A does not allow a foreign tax credit for this tax as the income of the branch for Country A purposes is zero (after application of carry-forward losses).

Adjustments made at time payment is set off against non-dual inclusion income

16. In order to defer the adjustment under Recommendation 4 until such time as the expenditure is actually set off against dual inclusion income, the taxpayer may need to maintain two memorandum accounts to keep track of:

   a. The amount of the potential adjustment that could be made under Recommendation 4. This memorandum account is similar to that recorded on the second line of Table B.10.4 and reflects the extent to which double deductions have exceeded dual inclusion income in each period.

   b. The change in the amount of unused loss in the branch jurisdiction. This memorandum account adjusts (up or down) the amount in the first account by reference to the change in the amount of the unused loss in the counterparty jurisdiction. This memorandum account measures the change in the carry-forward loss amount recorded in line 2 of Table B.10.3.

17. Table B.10.5 sets out the required adjustment under the primary rule where the adjustment is deferred until the double deduction is actually set off against non-dual inclusion income in Country B.

Table B.10.5. **Adjustment under Recommendation 4**

	Year 1	Year 2	Year 3	Total
Net income under Country A law	400	600	1 100	2 400
Adjustment under Rec. 4	400	200	(300)	
Change in loss carry forward under Country B law	(200)	100	100	
	600	900	900	
Tax under Country A law	(180)	(270)	(270)	(720)
Direct foreign tax credit	0	0	0	0
Total tax paid				(720)

Example 11

Imported mismatch

Facts

1. This example is based on the one set out in Figure 5 of this report. In this example A Co supplies services to a related company (C Co) through a branch located in Country B. The services supplied by the branch exploit underlying intangibles owned by A Co. Country B attributes the ownership of those intangibles to the head office and treats the branch as making a corresponding arm's length payment to compensate A Co for the use of those intangibles.

2. This deemed payment is deductible under Country B law but is not recognised under Country A law (because Country A attributes the ownership of the intangibles to the branch). Meanwhile, the services income received by the branch is exempt from taxation under Country A law due to an exemption or exclusion for branch income in Country A. It is assumed that there is no rule in either Country A or B addressing the mismatch in tax outcomes arising from the notional payment.

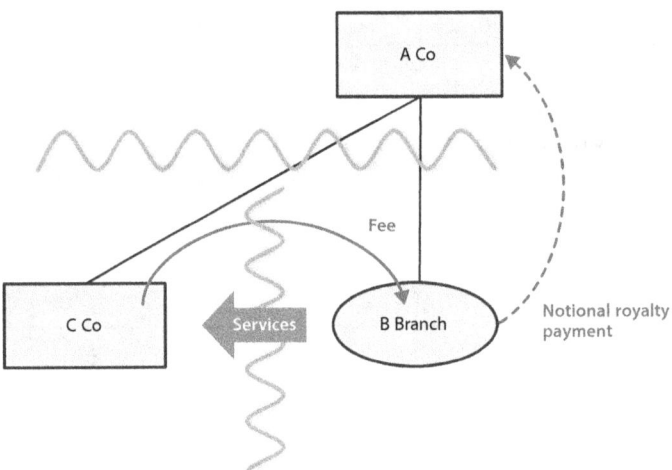

3. As a consequence, the (deductible) service fee paid by C Co (which is treated as exempt under Country A law) is offset against a deduction under a branch mismatch arrangement resulting in an indirect D/NI outcome.

Question

4. How does the imported mismatch rule in Recommendation 5 apply to neutralise the mismatch in tax outcomes arising from this structure?

Answer

5. The services fee paid by C Co will be subject to adjustment under Recommendation 5 to the extent the income from the payment is set off against a deduction under a branch mismatch arrangement. Recommendation 5 will not apply, however, if the income of Country B Branch was taxable under Country A law so that the service fee paid to Country B Branch was treated as dual inclusion income by Country B. In such a case, the offset of the service fee against the deemed branch payment would not give rise to a branch mismatch, and there would therefore be no adjustment required under the imported branch mismatch rule.

Analysis

Services fee is subject to adjustment under Recommendation 5

6. An imported branch mismatch is a transaction or series of transactions that is entered into between members of a controlled group that directly or indirectly funds deductible expenditure under a branch mismatch arrangement.

7. In this case, the deemed royalty payment made by the Country B Branch to its head office is a branch mismatch payment under Recommendation 3 and the services fee paid by C Co to B Co is a deductible payment that directly funds that deductible expenditure under that branch mismatch arrangement. The arrangement (including the branch mismatch and the payment by C Co) has been entered into between members of the same control group and accordingly the payment of the services fee will be subject to adjustment under Recommendation 5.1.

No imported mismatch if income of the branch is taxable under the laws of the residence jurisdiction

8. A payment that is set off against a deduction under a deemed branch payment should not be treated as having funded expenditure under an imported mismatch arrangement if that payment gives rise to dual inclusion income. Accordingly, if the Country B Branch was a taxable branch (so that the service fee paid to Country B Branch was included in income in both Countries A and B) then the payment would not be treated as funding expenditure under an imported branch mismatch and there would no adjustment required under Recommendation 5.

ORGANISATION FOR ECONOMIC CO-OPERATION AND DEVELOPMENT

The OECD is a unique forum where governments work together to address the economic, social and environmental challenges of globalisation. The OECD is also at the forefront of efforts to understand and to help governments respond to new developments and concerns, such as corporate governance, the information economy and the challenges of an ageing population. The Organisation provides a setting where governments can compare policy experiences, seek answers to common problems, identify good practice and work to co-ordinate domestic and international policies.

The OECD member countries are: Australia, Austria, Belgium, Canada, Chile, the Czech Republic, Denmark, Estonia, Finland, France, Germany, Greece, Hungary, Iceland, Ireland, Israel, Italy, Japan, Korea, Latvia, Luxembourg, Mexico, the Netherlands, New Zealand, Norway, Poland, Portugal, the Slovak Republic, Slovenia, Spain, Sweden, Switzerland, Turkey, the United Kingdom and the United States. The European Union takes part in the work of the OECD.

OECD Publishing disseminates widely the results of the Organisation's statistics gathering and research on economic, social and environmental issues, as well as the conventions, guidelines and standards agreed by its members.

www.ingramcontent.com/pod-product-compliance
Lightning Source LLC
Chambersburg PA
CBHW082346220526

45470CB00008B/2661